Soonish

Soonish

Ten Emerging Technologies That'll Improve
and/or Ruin *Everything*

KELLY AND
ZACH WEINERSMITH

PENGUIN PRESS
NEW YORK
2017

PENGUIN PRESS
An imprint of Penguin Random House LLC
375 Hudson Street
New York, New York 10014
penguin.com

LIBRARY OF CONGRESS CATALOGING-IN-PUBLICATION DATA
Names: Weinersmith, Kelly, author. | Weiner, Zach, 1982– author.
Title: Soonish : emerging technologies that'll improve and/or ruin everything /
Kelly and Zach Weinersmith.
Description: New York : Penguin Press, 2017. | Includes bibliographical
references and index.
Identifiers: LCCN 2017008654 (print) | LCCN 2017016790 (ebook) |
ISBN 9780399563836 (ebook) | ISBN 9780399563829 (hardcover : alk. paper)
Subjects: LCSH: Technological forecasting–Popular works. | Technological
forecasting–Pictorial works. | Technological forecasting–Humor.
Classification: LCC T174 (ebook) | LCC T174 .W543 2017 (print) | DDC
601/.12–dc23
LC record available at https://lccn.loc.gov/2017008654

Printed in China
1   3   5   7   9   10   8   6   4   2

DESIGNED BY AMANDA DEWEY

Dedicated to our parents,

Patricia and Carl Smith

and

Phyllis and Martin Weiner,

without whom this book could never have been completed.
You fed us, cared for us when we were sick,
watched Ada when we couldn't,
and made sure we came up for air now and then.
We will always appreciate what you did to make our dream real.
This book is as much yours as ours.[1]

---

1. Of course, we're keeping the check to ourselves. But the sentiment is there.

CONTENTS

1. Introduction: *Soonish. Emphasis on the Ish*   *1*

SECTION 1
The Universe, Soonish

2. CHEAP ACCESS TO SPACE: *The Final Frontier Is Too Damn Expensive*   *13*

3. ASTEROID MINING: *Rummaging Through the Solar System's Junkyard*   *52*

SECTION 2
Stuff, Soonish

4. FUSION POWER: *It Powers the Sun, and That's Nice, but Can It Run My Toaster?*   *73*

5. PROGRAMMABLE MATTER: *What If All of Your Stuff Could Be Any of Your Stuff?*   *101*

6. ROBOTIC CONSTRUCTION: *Build Me a Rumpus Room, Metal Servant!* *134*

7. AUGMENTED REALITY: *An Alternative to Fixing Reality* *164*

8. SYNTHETIC BIOLOGY: *Kind of Like Frankenstein, Except the Monster Spends the Whole Book Dutifully Making Medicine and Industrial Inputs* *190*

SECTION 3

You, Soonish

9. PRECISION MEDICINE: *Everything That's Wrong with You In Particular—a Statistical Approach* *229*

10. BIOPRINTING: *Why Stop at Seven Margaritas When You Can Just Print a New Liver?* *257*

11. BRAIN-COMPUTER INTERFACES: *Because After Four Billion Years of Evolution You Still Can't Remember Where You Put Your Keys* *282*

12. CONCLUSION: *Less Soonisher, or The Graveyard of Lost Chapters* *318*

Acknowledgments *337*
Bibliography *339*
Index *351*

Soonish

1.

Introduction

Soonish. Emphasis on the Ish.

This is one of those books where we predict the future.

Fortunately, predicting the future is pretty easy. People do it all the time. Getting your prediction right is a bit harder, but honestly, does anyone really care?

There was a study in 2011 called "Are Talking Heads Blowing Hot Air,"[1] in which the predictive abilities of twenty-six pundits were assessed. Predictive powers ranged from mostly right to usually wrong.[2]

For most people, the pleasure of reading this study was the discovery that certain individuals were not *just* intolerable morons, but *statistically* intolerable morons. From our perspective as pop science writers, there was an even more exciting result: Regardless of their predictive prowess, *all these people still have jobs.* In fact, a lot of the worst predictors were the most prominent public figures.

If there really is no relationship between predictive ability and having a successful career, we've put ourselves in an excellent position. After all, those pundits were just trying to predict what will happen in the short term among a

---

1. By a group of public policy students at Hamilton College. It is, in fact, a small sample set. But given that it confirms our biases, we choose to believe it.
2. Fun fact: Having a degree in law correlated with being worse at prediction.

small number of squabbling political actors. They weren't trying to decide if we'll have an elevator to space in fifty years or if we'll be uploading our brains to the cloud soon,[3] or if machines will print us new livers and kidneys and hearts, or if hospitals will use tiny, swimming robots to cure diseases.

Frankly, it's really freakin' hard to tell you whether any of the technologies in this book will be realized in their fullest form in any particular time frame. New technology is not simply the slow accumulation of better and better things. The big discontinuous leaps, like the laser and the computer, often depend on unrelated developments in different fields. And even if those big discoveries are made, it's not always clear that a particular technology will find a market. Yes, time travelers from the year 1920, we have flying cars. No, nobody wants them. They're the chessboxing[4] of vehicles—amusing to see once in a while, but most of the time, you'd rather have the two parts separate.

Given that any prediction we give you is likely to be not only wrong, but stupid, we've decided to employ some strategies we learned while reading other books where the authors envision the future.

First, a few preliminary predictions:

We predict that computers will get faster. We predict screens will get higher resolution. We predict gene sequencing will get cheaper. We predict the sky will remain blue, puppies will remain cute, pie will remain tasty, cows will continue mooing, and decorative hand towels will continue to make sense only to your mom.

---

3. Apple iCloud if this book does well. Amazon Cloud if this book does poorly.
4. This is a real sport, which is unsurprisingly popular in Russia. You alternate rounds of chess and boxing until you lose at one of them.

We urge you to check back in a few years to grade our accuracy. Please note that we specified no time frame, so your grading options are either "correct" or "not *not* correct."

Now that we've made the first round of predictions, we're prepared to make a few more. We predict reusable rockets will lower the cost of rocket launches by 30–50% in the next twenty years. We predict it will be possible to diagnose most cancers with a blood test in the next thirty years. We predict that nano-bio-machines will cure most genetic disorders in the next fifty years.

Okay, that's a total of eleven predictions. We believe that if we get eight out of eleven, we should be considered geniuses. Oh, and if any of the first set comes true, you can write clever news articles with titles like "COUPLE WHO PREDICTED THE FUTURE OF GENE SEQUENCING SAY SPACEFARING WILL BE CHEAP IN NEAR FUTURE."

Predicting the future accurately is hard. Really hard.

New technologies are almost never the work of isolated geniuses with a neat idea. As time goes on, this is more and more true. A given future technology may need any number of intermediate technologies to develop beforehand, and many of them may appear to be irrelevant when they are first discovered.

One recently developed device we discuss in the book is called a superconducting quantum interference device, aka a SQUID. This very sensitive device detects subtle magnetic fields in the brain, which is one way to analyze people's thought patterns without drilling holes in their skulls.

How did we get this thing?

Well, a superconductor is any material that conducts electricity without losing any electricity on the way. This is different from a regular old conductor (like a copper wire), which transmits electricity pretty well, but loses some en route.

We have superconductors because about two hundred years ago, Michael Faraday was making some glassware and accidentally turned a gas into a liquid by trapping it under pressure in a glass tube. There wasn't TV back then, so a bunch of Victorians got really excited about the idea of liquefying gasses.

As it turns out, it's easier to liquefy gasses by getting them really cold rather than getting them really pressurized. This insight led scientists to develop advanced refrigeration technology, which allowed them to liquefy stubbornly gassy elements, like hydrogen and helium. And once you have liquid hydrogen or helium, you can use them to cool down just about anything you like.

Helium, for example, is at about -450 degrees Fahrenheit when in liquid form. If you pour it onto just about anything, the liquid helium turns into a gas and takes heat away with it, until the thing you're cooling is also about -450 degrees.[5]

Eventually scientists wondered about what happens to conductors when you

---

5. To understand why, think of it like pouring cold water on a hot pan. The pan transfers a bunch of its heat to the water and thus cools down. You can cool it down faster by pouring off the water and getting new cold water. That cold water is something like 50 degrees Fahrenheit, so you can keep cooling the pan until it gets down to 50 degrees Fahrenheit. After that, the water is the same temperature as the pan, so the heat can't just go from one thing to another. It'd be sort of like trying to dry off with a towel that's just as wet as you are. You can't get dryer without a dryer towel, and you can't get colder without a colder cooling liquid.

get them *really* cold. Conductors tend to get better at what they do as they cool down. In simple terms, this is because conductors are sort of like pipes for electrons, but they're not perfect. In a copper wire, for example, the copper atoms get in the way of electron motion.

What we call "heat" is really just rapid wobbling at an atomic level. When you heat (aka wobble) atoms in copper wire, they are more likely to block electrons from moving downstream, in the same way it's harder to get down the street if the guy in front of you keeps changing lanes over and over. At the level of atoms, wobbling (aka heat) means the electrons are more likely to bump into the copper atoms, increasing the wobble still more. This is why your laptop charger gets really hot after you use it for a while.

When you put that liquid helium on the conductor, the wobble energy in the copper atoms is transferred to the helium atoms, which then fly away. Now your copper atoms are less wobbly and your electrons experience a lot less resistance. The colder they get, the easier it is for electrons to flow.

Back then there was a debate about what would happen when you got toward *zero* wobble. Some thought conductance would cease because at that temperature motion should be impossible, even for electrons. Some thought conductance would get very good, but nothing special would happen.

So researchers started to pour their ultracold gasses onto metal elements. It turned out, bizarrely, that some metals became *perfect* conductors (aka superconductors) when they reached a certain very low temperature. If you kept the metal cold enough to superconduct, you could put electric current in a loop, and it would just keep looping forever. This may sound like a cute science fun fact, but it leads to all sorts of weirdness! That looping current would generate a magnetic field. And that means you could turn these cold metals into permanent magnets, whose magnetic strength was determined by how much current you added.

Later, in the 1960s, a guy named Brian Josephson (who got a Nobel Prize, but now spends his days defending magic nonsense like cold fusion and "water memory" at Cambridge) discovered an arrangement of superconductors that allows you to detect tiny variations in magnetic fields. This device, called a Josephson junction, eventually allowed for the development of the SQUID.

Now then. Consider this: If someone came to you two hundred years ago and asked how we might build a device to scan people's brain patterns, would your immediate response be, "Well, first we need to trap some gas in a glass tube"?

We suspect not. In fact, even the last big technical step—the Josephson junction, which again was discovered by a man who thinks it's possible that *water remembers what you put in it*—was considered theoretically impossible when it was first proposed. Its behavior was explained later, using a theoretical framework developed long after Michael Faraday was dead.

The contingent nature of technological development is why we don't have a lunar base, even though we thought we would by now, but we do have pocket-sized supercomputers, which few people saw coming.[6]

The same difficulty holds for all the technologies in this book: Whether we can build an elevator to space may depend on how good chemists get at arranging carbon atoms into little straws. Whether we can make matter that assumes any shape we tell it to may depend on how well we understand termite behavior. Whether we can build medical nanobots may depend on how well we understand origami. Or maybe none of that stuff will end up mattering in the end. There is nothing about history that necessarily had to be as it was.

We now know that the ancient Greeks could create complex gear systems, but never constructed an advanced clock. The ancient Alexandrians had a rudimentary steam engine but never designed a train. The ancient Egyptians invented the folding stool four thousand years ago, but never built an IKEA.

---

6. This sort of thing sometimes causes people undue distress, as in the recent *MIT Technology Review* cover, featuring moonwalker Buzz Aldrin with the headline "YOU PROMISED ME MARS COLONIES. INSTEAD I GOT FACEBOOK." But, in fairness, a Mars colony would cost a few trillion dollars, while Facebook is free. And, it's worth noting that the choice of Facebook is a bit crafty. Imagine if they'd picked Wikipedia: "YOU PROMISED ME MARS COLONIES, AND ALL I GOT WAS ALL OF HUMAN KNOWLEDGE INDEXED AND AVAILABLE TO EVERYONE ON EARTH FOR FREE."

All this is to say—we don't know when any of this stuff is going to happen.

So why write this book? Because there are *amazing things* happening all over the place every day, all the time, and most people aren't aware of them. There are also people who become cynical because they thought we'd have fusion power or weekend trips to Venus by now. This disappointment is not always due to scientists who overpromise the future; often books like this one omit the economic and technical challenges that stand between us and the future as depicted in fiction.

We don't know why these challenges are so often left out of books. Would the story of Apollo 11 be better if getting to the moon were easy? To our way of thinking, part of what makes the idea of a brain-computer interface so exciting is that right now we have almost no clue how to decode thoughts. There is an unlimited frontier of questions to be asked, discoveries to be made, glory to be won, and heroes to be garlanded.

We picked out ten different emerging fields to explore with you, and we ordered them roughly from large to small, moving from outer space, to giant experimental power plants, to new ways to build things and experience the world, to the human body, finally all the way down to your brain. No offense.

Our guiding principle for each of these chapters was this: If you were sitting at a bar, and someone asked you, "Hey, what's the deal on nuclear fusion power," what would be the best answer possible? We were told we don't know what bars are like, *but the point is* that each chapter will tell you what the technology is, where it is right this second, the challenges to its realization, the ways it might make everything terrible, and the ways it might make things wonderful.

To us, scientific progress isn't just exciting because it does new things for us. Knowing how damn hard it would be to mine an asteroid or build a house with a robot swarm makes those things *more* interesting. And it means that when these things finally *do* happen[7] you'll understand exactly how exciting it is.

You'll also understand a bit about the strange detours and blind alleys science and technology take. At the end of most chapters, we provide a nota bene on some nugget of weirdness (or grossness or awesomeness) we unearthed. Sometimes these sections are directly related to their chapters, and sometimes they're just weird, weird things we bumped into while doing our research. Like, really weird. Like, octopus-made-of-cornbread weird.

For all these chapters, we had to read a lot of technical books and papers and we had to talk to a lot of mildly crazy people. Some were crazier than others, and generally they were our favorites. The one unifying experience in all our

---

7. Even as we wrote this book, two technologies in it took a major leap. We had to amend our cheap access to space chapter after SpaceX repeatedly landed booster stages of its Falcon 9 rocket, and we had to amend our augmented reality chapter because people will not stop talking about Pokémon GO.

research was that on every single topic all of our preconceptions were crushed. In every case, as we researched we discovered that we not only hadn't understood the technology itself, but we hadn't understood what was holding it back. Often what seemed complicated was easy, but what seemed easy was complicated.

New technologies are beautiful things, but just like with Michelangelo's *Pietà* or Rodin's *Le Penseur*, it's usually an unholy pain in the ass to make them. We want you not just to understand what a technology is, but to understand why the future so stubbornly resists our best efforts.

Kelly and Zach Weinersmith
Weinersmith Manor, September 2016

P.S. We also want you to know about this one experiment in which undergrads were forced to breathe through one nostril, then take exams. It's kinda relevant. We promise.

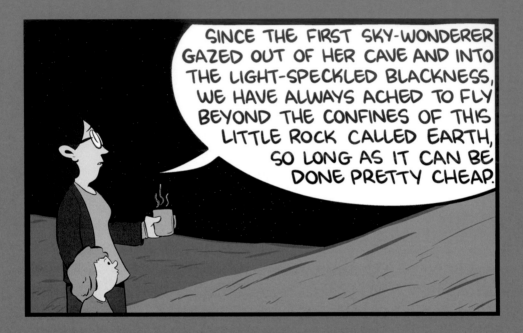

SECTION 1

The Universe, Soonish

2.

Cheap Access to Space

The Final Frontier Is Too Damn Expensive

Up, up the long, delirious, burning blue
I've topped the wind-swept heights with easy grace.
Where never lark, or even eagle flew—
And, while with silent, lifting mind I've trod
The high untrespassed sanctity of space,
—Put out my hand, and touched the face of God.

· John Gillespie Magee, Jr., "High Flight," 1941

One thing you'll immediately notice about this poem is that he never once talks about price. This is the kind of glaring technical omission often made in poetry, so we're adding one more couplet:

And when I asked what space was priced,
I turned around, 'cause HOLY CHRIST!

Right this second, it runs you about $10,000 to send a pound to space.[1] That's about $2,500 per cheeseburger.

---

1. This number actually varies a lot, and depends on things like the country from which you're flying, the company you go with, where you're going, and the size of the space vehicle that is transporting the stuff. We're using $10,000 per pound as a ballpark figure throughout the book. Adding or subtracting $9,000 from that figure encompasses all estimates we've encountered while researching this topic.

This is why human beings have only been to the moon's surface half a dozen times, and it's why our moon vehicles were paper thin in places. The fact that in 2017 we have a space travel paradigm that would've disappointed all the hopes of 1969 is not due to a lack of engineering or scientific genius. It's because the cost of the way we get to space has remained stubbornly high. If we could dramatically reduce the cost, we would have better space science, better communication systems, access to off-planet resources, better ability to control our climate, and best of all, the solar system would open up for exploration and settlement.

To understand why it's currently so expensive to get stuff up to space, you need to understand what you're looking at when you see a rocket.

A rocket is essentially a tube of explosive propellant with a liiiittle bit of cargo on top. For a typical mission going to Low Earth Orbit (LEO; about 300 miles high, and where most launches go), by mass you're looking at 80% fuel, 16% the rocket itself, and 4% cargo (4% is actually on the high end, and if you're going farther out, it gets closer to 1 or 2%).

But when you look at cost, things are inverted. The propellant is a negligible component of price—it's gonna run you a mere few hundred thousand dollars. So most of the cost is taken up by the rocket itself, which is almost always discarded after use.

In sum, launching rockets is really expensive and most of the space onboard is taken up by propellant. This leaves two ways we can try to drastically lower the cost to make space access cheap:

1. Recover the launch vehicle.
2. Use less propellant.

Vehicle recovery suddenly became a reality in 2015, which we'll get into in the section on reusable rockets. But the basic idea is pretty simple—you can save money if you don't junk your vehicle after one use.

Using less propellant is a little trickier, even though propellant is 80% of a spacecraft's starting mass. To understand why, consider a situation where you have to drive from Russia to South Africa and back again. You're offered two ways to get your fuel:

1. Gas up at stations along the way.
2. Take all the fuel you'll need for the whole trip and drag it along with you.

Of course, you'd rather use option 1. But consider why, in particular.

A car is just a machine that converts fuel into forward motion. If your car is really heavy, it takes more fuel to get a certain amount of forward motion. If you gas up regularly, most of your weight is the car and not the fuel. This means the fuel the engine is using right this second is supplying forward motion mostly to the vehicle (and you, and your luggage) and not to the fuel in the tank.

In the case of option 2, you're dragging an enormous tanker. The weight of fuel is probably far, far higher than the weight of the car itself. Especially at the beginning, you're using most of the energy derived from the fuel just to move the fuel itself. So *most of the fuel's energy goes to moving other fuel.*

The result? The total amount of fuel you need is far higher in case 2 than in case 1. Your little caravan, just like all space rockets, is mostly made of fuel, not of vehicle or cargo.

Unfortunately, it's a little hard to build gas stations for rockets. So without a major change, we're stuck in scenario 2 when it comes to space travel.

All of this sets up some very tantalizing math. If you could make the launch vehicle recoverable, you could potentially eliminate 90% of the cost of space launch. Or, if you could use just three quarters as much fuel, you'd be able to fit six times as much cargo,[2] instantly dividing the cost per pound by six.

The hard thing here is that you're fighting fundamental physics. The cheapest orbit available is LEO. People often think that "orbit" means there's no gravity. This is incorrect. In fact, the International Space Station (which is in

---

2. Fuel is 80%. Three quarters of that is 60%. That frees up 20%. But cargo originally only took up 4%. So you have increased from 4% cargo to 24% cargo!

LEO right now) is usually around 250 miles high and experiences about 90% of the gravity you experience on Earth. So why do the astronauts float around like there's no gravity? Because they are going really, really, really fast. About 5 miles per *second*. Although they are pulled toward the Earth all the time, they always "miss" it.

Think of it like this: Imagine you fire a cannonball from the top of a tower. If you fire it softly, the ball will go a little ways then fall to the ground. If you fire it incredibly fast, it will just fly off into space. But between falling right down and going off into space, there are a lot of intermediate regimes. For a given height, there is some speed that is slow enough that it can't leave Earth, but fast enough that you'll never plop to the ground. If you were riding that cannonball, you'd be falling, because gravity is tugging you down. At the same time, because you're going so fast, you'd be able to see Earth's curve. As you move from a point on the globe in a straight line, Earth curves down and away from you, increasing your distance from the surface. At this particular speed, you have two balanced effects: Gravity wants you down low, but your speed keeps you up high. So you just keep going around and around and around. You "orbit."

Even though LEO is the cheapest orbit to achieve, it's still pretty expensive to get there. Getting a big hunk of metal to 5 miles per second is not an easy task. If we ever want spaceships that look like the ones in movies instead of giant tin cans wrapped in foil, we're going to need a cheaper way.

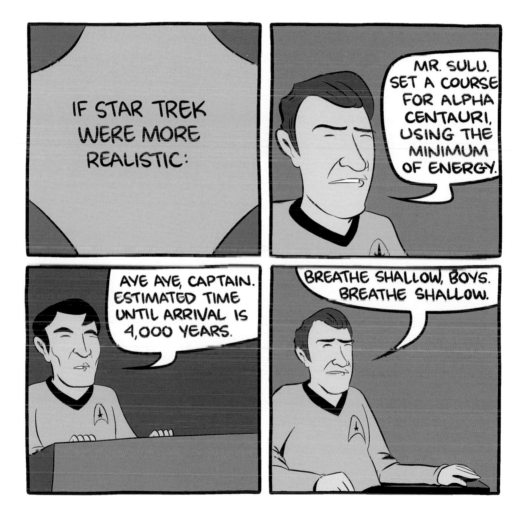

Where Are We Now?

Method 1: Reusable Rockets

Reusable rockets are the best bet for cheaper spaceflight in the short term. They are traditional rockets, but rather than falling into the ocean as they do now, they fall to Earth and land after they finish the mission. This doesn't fix the problem that the rocket only holds 4% cargo, but it potentially drives the cost way down.

There are a few difficulties with this approach, though. You have to keep extra propellant onboard for the landing phase, which lowers efficiency. You want to carry the smallest amount of extra propellant possible, but this makes the landing phase very hard.

A very serious issue is that nobody yet knows what it'll cost to refurbish a used rocket. This thing has gone to space, man. You can't just put a spit shine on it and put it back on the launchpad.

The U.S. Space Shuttle, which was designed to be a reusable launch vehicle, ended up being more costly than a regular rocket precisely because refurbishing was so expensive. There's an ongoing argument over whose fault this was—the engineers, Congress, the Air Force, a risk-averse public, and more—but the bottom line is that the program was largely done in by the cost of getting the Shuttle launch ready again after a flight. This is why, when lots of people were sad about the Shuttle retiring, a lot of space nerds were glad to see it go.

But there is reason to hope that a better reusable launch vehicle can be created. As we were writing this chapter, SpaceX became the first company to successfully put cargo into space, then land part of its rocket.[3]

If it really can bring the price down, this may prove to be the biggest development in space travel in a generation. As we were watching a launch, a reader

---

3. Actual rockets have several sections called "stages." Once you use up a stage, it's deadweight that's slowing you down. So you dump it. SpaceX recovered the first booster stage, which is the largest.

of ours tweeted that although he had witnessed the moon landing as a young boy, he found the reusable rocket even more exciting. It sounds crazy, but he's got a point—the moon landing was certainly the greater technical feat, but it was done at a cost that more or less guaranteed it couldn't become commonplace. Exactly how much the cost can be dropped is a matter of debate. Elon Musk apparently claimed he could eventually get the cost down by a factor of 100. In the more near term, SpaceX's president Gwynne Shotwell said their current Falcon 9 should be able to offer a 30% discount. But even if reusable rockets only mean a small price drop now, they may yet represent a path to greater future savings. The road to Mars may be paved with small discounts.

Method 2: Air-Breathing Rockets and Spaceplanes

Airplanes already go really high. Can't we just have them go a bit higher so they get to space?

No. Why would you even ask that? Jesus.

If you want to put a satellite in orbit, the hard part is not going really high. The hard part is going really fast. That takes a lot of propellant. But using a spaceplane might allow a serious reduction. To understand why, you have to understand what propellant is.

If you refer to propellant as "fuel," a NASA engineer will beat you with a TI-83.[4] Propellant is actually a combination of two things: fuel and oxidizer. When you want a combustion reaction, you need three things: fuel, oxidizer, and energy. For example, when you light a campfire, the fuel is wood, the oxidizer is (you guessed it) oxygen, and the energy is a lit match.

In a rocket, you carry both fuel and oxidizer inside the ship. The actual ratio of oxidizer to fuel varies by rocket and mission, but generally speaking the *majority* of the propellant's mass is oxidizer. The oxidizer is often just liquid oxygen.[5] Why carry all that liquid oxygen when the rocket is literally surrounded by oxygen for much of its trip?

The short version is that we're keeping it simple. A rocket is a brute force way to get to space. You put everything you need in a big tube and blast your way skyward. With an airplane, you might be able to improve your efficiency by getting your oxidizer from the air rather than carrying it with you, but you're adding a lot more complexity to an already complicated machine.

The big problem for a spaceplane is that you need multiple types of engines to handle all the different speeds and conditions you encounter en route to space. Here's why:

Most airplanes today use a turbofan engine. They're a bit complicated, but the basic mechanism is simple. Fans suck air into a chamber. The air is compressed, so you have a lot of oxygen (your oxidizer!) in a small space. Fuel is injected and ignited. The result is hot compressed air that you channel out the back as you suck in more air. Now, you've got high-pressure air behind the

---

4. We asked Twitter what a rocket engineer is most likely to use to beat someone to death, and the most frequent suggestions were a TI-83, a TI-89, a TI-30X, a slide rule, or just a reasonably good laptop with MATLAB installed.
5. Delightfully abbreviated as "LOX."

engine and comparatively low pressure in front of the engine. So you go forward.

Turbofans start having trouble when you get toward the speed of sound, at about 767 miles per hour,[6] also known as Mach 1. At the speed of sound, the air can't get around the plane as fast as it builds up. This creates problems if your front intake is a fan.

One solution to getting over this hump is what's called an afterburner. An afterburner takes leftover oxygen at the back of your turbofan, throws more fuel at it, and ignites it. In short, you make a little ongoing fuel explosion at the back of your plane. By this means you can get toward Mach 1.5, though not terribly efficiently. But once you're at Mach 1.5, you can use a different type of engine called a ramjet.

A ramjet is an incredibly simple machine, but it's not necessarily easy to make. Basically, you have a turbofan engine minus all moving parts, including the fan. You don't need a fan to compress the air because your high speed is doing it for you. You fly fast, and air crams into a chamber where it slows down as you add fuel and ignite. The downside here is that because speed itself is your compressor, you can't *start* with a ramjet. You can only use a ramjet once you're going about 1100 miles per hour. So, for example, on an SR-71 spy plane, you have a turbofan that changes its shape to behave like a ramjet once you get to the right speed.

Once you get really, really fast (but still not fast enough to stay in LEO), you need a supersonic ramjet, or "scramjet." A scramjet is an even simpler machine that is even harder to build. Basically, supersonic air comes in and, along with fuel, gets ignited directly, without ever slowing it down. You can do this because the oxygen is coming so fast, there's enough to get a combustion reaction going without compression. But it's not easy to, so to speak, light a candle in supersonic wind. Scramjets are still experimental, but after about 4500 miles per hour[7] they become the most efficient way to go. In theory, they can take you all the way up to Mach 25, which is orbital speed. There have been a number of

---

6. This number can change a bit, as the speed of sound depends on things like temperature and elevation.
7. That's Tokyo to London in about two hours, if you don't count the acceleration time.

scramjet programs, most of them military, and all have met with only limited success. None of them have yet come anywhere near orbital velocity.

An ideal spaceplane should be able to make use of all these engine types in sequence to get to space. Once in space, where there is no available oxygen, you will probably switch over to a traditional rocket propellant method. But by using oxygen from the air instead of an onboard tank, you can cut down fuel use enough that you might be able to carry ten times more cargo.

Oh, and since it's a plane, it can just land afterward. If this can be done repeatedly, without too much damage, you've solved the vehicle loss problem *and* the fuel efficiency problem.

The hard part is that all these machines have to work under extreme conditions. The conditions a scramjet is optimized for are *so extreme* that they're expensive just to simulate down here on Earth.

A British firm called Reaction Engines is working on a vehicle called Skylon, which uses an engine called SABRE, for Synergetic Air-Breathing Rocket Engine. We're guessing they came up with the "ABRE" part quickly, then spent a few days deciding on an *S*. In short, it's a rocket, but it takes in ambient oxygen as part of its thrust reaction. Their engine is designed to efficiently switch from a turbofan to a ramjet to a rocket. Presumably they aren't doing a scramjet phase because, well, nobody really knows how to *do* the scramjet phase.

It's an expensive, complicated endeavor, but they do have substantial funding from the European Space Agency[8] and the British government. If things go well, they hope to field one of these advanced planes in the next decade.

For all the downsides of rockets, they have the virtue of simplicity. An old-fashioned rocket works just fine in low speed or high speed, in thick atmosphere, thin atmosphere, and no atmosphere. So, hey, how about we try something even more old-fashioned?

---

8. In case you're wondering, Brexit shouldn't interfere with this program. The European Space Agency works closely with the European Union, but it already has non–European Union members (Norway and Switzerland) and is not controlled by the European Union.

Method 3: Giant Giant Giant Enormous Mega-Superguns

One way to save on rocket fuel is not to use any. Earlier we discussed how rockets are encumbered by the need to use propellant to accelerate propellant. What if instead of running a relatively slow controlled burn all the way to space, we had one giant boom down here on the ground? Sure, you have to use a lot of explosives overall, but none of those explosions are used to lift more explosives. This should save a lot of overall energy.

Mind you, it won't be cheap. It's a cannon that would probably be thousands of feet long, with a barrel on the order of 10 feet in diameter, packed with literally tons of explosives. But there are advantages: no discarded parts, no using fuel to carry fuel, and pretty much every bit of each shot actually goes to space.

It's not quite as crazy as it sounds, and there have been at least two well-funded government projects to explore this method, one of which we discuss in the nota bene for this chapter. But there are two major drawbacks:

First, every time you fire, you have to create an enormous explosion. So if you want to use this thing repeatedly without too much expense, you need some kind of chamber that can withstand several tons of explosive material being detonated regularly.

Second, getting shot out of a cannon isn't very fun. Well actually, if you were shot out of a space cannon, it wouldn't be fun or not fun. You'd just be splatted.

It's not the speed that kills you. It's the acceleration—the *change* in speed.

When you go up in an elevator, you feel as if you're getting squashed. That's a slight acceleration. By comparison, on a roller coaster, you may feel as much as five times more acceleration. With training, humans can endure about ten or twenty times the elevator without passing out. Much beyond that and you might die. Why? Well, when you accelerate in a car, notice how the water in a cup rushes back and stays back until you stop accelerating. Imagine the cup is your body and the water is your blood. Oh, and instead of 0 to 60 in 10 seconds, you're going from 0 to 17,000.[9]

For an explosion-based space cannon, you're talking around 5,000 to 10,000 times the elevator. Nothing squishy is going to space in a cannon, including squishy little you.

This may not be as bad as it seems. You could still send "hardened" payloads, like specially designed electronics. You could also send all sorts of raw material—metals, plastics, fuel, water, beef jerky. In fact, one idea is to have a sort of orbiting gas station that just receives fuel payloads from a gun.

By itself, a space gun is not a great route to space exploration. But if you coupled a space gun with an orbiting factory in space, we might be in business. The idea here would be to fire raw materials up to your orbiting factory, build gigantic spacecraft at the factory, and then take off from the factory to go explore space. For annoyingly delicate payloads, like humans, you'd still need a wussier form of launch, like rockets. But on a big space mission that's already in space, most of what you're toting is metal, plastic, and supplies for the delicate meatbags within. All these things can be "ruggedized" and shot to orbit.

Another option is to have a gun that speeds up slowly enough that the cargo experiences a more human-friendly level of acceleration.

For instance, you could have a sequence of explosions, spreading the acceleration over time. The downside is that you're taking an expensive and difficult system and making it more expensive and more difficult. Now you've got dozens of explosions instead of one, which means a longer barrel and a lot more potential for error.

Another option is to have an electromagnetic railgun. Basically, you start with a magnetically levitated train. These "MagLev" trains float on a magnetic field, which is important because with conventional rail, beyond a certain velocity you'll start bending and even melting the tracks. You put this vehicle and its

---

9. So why don't you die when you accelerate in the car? You're not accelerating fast enough for a long enough time. Plus, your body is a lot closer to a sponge than an open cup. Your circulatory system resists the change in speed just fine. If the acceleration were higher and sustained, you'd be a lot more like the cup. Of course, you can also die from sudden rapid speed changes, also known as "car crashes."

track in an airless tube that is about 100 miles long. Then you keep using powerful magnetic fields to boost its speed. It's basically explosive speed boosts without the boom. The upside is the method is a lot cleaner and easier to reuse. The downside is the necessary materials—specifically the ultralong evacuated tube and train system—would be much more expensive.

But this too has a problem: Even if you spread the acceleration out over time, the projectile at some point must exit the tube, going from an airless environment into the atmosphere at hypervelocity.

To understand what happens when the projectile exits the tube, consider this: Moving through air is the same as having air move past you. It's air particles thwacking against your body. The most extreme winds on Earth happen in tornados, and the fastest winds recorded are around 300 miles per hour. If you want to reach orbit at the right velocity, you need to be fifty to a hundred times faster than that as you fire out of the cannon.

At that speed, the air is fighting you so hard that it will literally ignite. So not only have you got a lot of air drag, you also have an explosion. Not great for cargo.

One way around this problem is to build the tunnel structure so high that the cargo doesn't leave the tube until it's in the thin upper atmosphere. The atmosphere gets less dense very rapidly once you get around 25 miles up. The issue with this approach is that we don't know how to build anything that stands 25 miles tall. The tallest structure humans have ever made is about half a mile from bottom to top,[10] and it's a skyscraper, not a launch track. Even if we knew how to build it, it would cost an insane amount of money.

But people are still trying to make gun methods work, and there are a couple variants on this concept, including two of the best-named ideas in this chapter: the "Slingatron" and the "rocket sled."

The Slingatron is a railgun on a spiral track. We talked to Jason Derleth from NASA Innovative Advanced Concepts (NIAC), which is a sort of asylum for people with really crazy space ideas that just might work. He told us, "The

---

10. The Burj Khalifa, in Dubai.

Slingatron is unfortunately highly unlikely to work. I really like it. I think that it's a brilliant idea, but what ends up happening is you have to put it at the top of Mount Everest for it to have even a chance, because it's fighting air resistance the entire time."

The rocket sled is basically another railgun, but instead of accelerating the projectile, it's accelerating a sled that carries a rocket. The sled goes really fast, getting you up to high speed into the thinner part of the atmosphere. Once you're up high, you start up the rocket. The extra speed and height get you a serious fuel savings. *Plus,* you've got a rocket sled.

All these methods could potentially be combined with a ramjet/scramjet system. Remember, those work once you're fast, and are necessarily designed to handle extreme conditions. But, as with all hybrid systems, you're developing something even more complicated and perhaps only getting a bit more efficiency.

Which leads us to this other concept we learned about from Mr. Derleth.

Method 3.5: Giant Giant Giant Enormous Mega-Super . . . Pogo Stick?

"One of the most interesting ideas that I heard was ludicrous on the face of it, okay? Really, really stupid . . . One person suggested, well, why don't we just put the shuttle on a pogo stick? I mean, an actual mechanical thing that you can press down, like a big spring, and give it a little bit more oomph at the beginning. It sounds so stupid, and yet, you probably could actually get another percent more payload if you did something like that. It's brilliant, it's really neat."

Method 4: Laser Ignition

Rockets basically work by firing hot stuff out the back. The hotter it fires, the more of a boost you get per volume of propellant. One way you might get things *really* hot would be to have a super-high-powered laser onboard to zap the fuel to extremes of heat. But this would weigh so much that it wouldn't be worth it.

So scientists had an idea that probably left astronauts unenthusiastic: Could we fire a laser right up the rear end of a flying rocket? When we mentioned this to European Space Agency's Michel van Pelt, he pointed out, "This may be just something to get used to. I mean, if you told people fifty, sixty years ago that you would go into orbit basically sitting on a pile of rocket fuel, basically a controlled explosion, that also probably doesn't sound too appealing either."

You could save a lot of fuel this way. In fact, one group suggested that with a powerful enough laser, you could use zero fuel up to the first 7 miles of atmosphere. You could gain speed simply by megaheating the air under the rocket. Once you get high enough, you have to use fuel, but thanks to the added laser heat you'd still need far less.

The problem? We're talking a *huge, HUGE* laser—in terms of power output, something on the order of 50,000 megawatts. That's roughly equivalent to the combined output of fifty nuclear reactors all at once. Mind you, you only have to fire the laser for about ten minutes. But even if that weren't an *insane* amount of energy, well, we don't even know how to build a laser that powerful. The most powerful lasers that can fire continuously are U.S. military weapons, and they top out around 1 megawatt. And they only fire for about a minute.

That said, if we *could* build giant megalasers, there might be an additional bonus for rocketry. One group at Brown recently suggested that a powerful laser could be used to reduce air drag by as much as 95%.

Imagine this: As you are being laser-blasted up, a second laser is being fired into the region ahead of you. This makes the air ahead of you less dense, so there's less to bump into. Now your astronauts might get a little antsy, since they're flying at well past the speed of sound with ultrapowerful lasers before and behind them, but you could solve this problem by just calling them cowards.

As a bonus, having a thin air patch with denser air around it can even help steer the rocket, for the same reason that if you're running through a bar, you will naturally move through the area with a less dense crowd.

One issue with all this, *if you're a wimp*, is that a 50,000-megawatt laser is an incredible weapon. Like, you could instantly incinerate just about anything from a long distance. This might make for some geopolitical headaches. But hey, maybe if we show other countries how cool our DOUBLE-LASER ROCKET is, they'll care less about the existential risk it poses to all nations on Earth. Or at least they'll say less about it.

Method 5: Start at High Altitude

Okay, so as we discussed, altitude isn't your main foe—speed is. However, starting at high altitude *does* mean you start at the thin part of the atmosphere. Once you're about 6 miles high, the air is 90% thinner. This is why airplanes spend so much fuel to get up high. Once you're at 38,000 feet (about 7 miles high), there's way less air drag.

We are going to focus on three proposals for starting up high: the rockoon, the stratospheric spaceport, and the aircraft-launched rocket.

A rockoon is just a rocket that gets floated up by a balloon, then hits the ignition once it's up high. It's really not a good way to go for a big rocket. Once a balloon is going up, you pretty much have no fine control over it. This is less than ideal when you're about to ignite a skyscraper-sized tube of propellant. Rockoons were tried back in the fifties, but were quickly abandoned as a method for space launch. Still, dedicated nerds occasionally fly them to get nifty photos and, we suspect, to use the word "rockoon."

But couldn't we have a stratospheric spaceport? Pretty please? Could we?

A modern rigid airship (picture a slightly more sleek zeppelin) can carry around 10 tons of cargo. A modern rocket, fully fueled, weighs about 500 tons.

As awesome as a fifty-blimp armada sounds, it's not going to be cheap or easy to maintain.

Alternatively, you could create the giant spaceport purely to hold up the launch end of the MagLev track described earlier. That way, you get a high-elevation exit from the launch tube without having to build a permanent 25-mile-high structure. But in order to do this, now you need an even larger floating structure, since it has to hold an enormous track.

Long story short, big airships are likely to be a bad way to go. We mention it mostly because it's the first idea most people have when they think about space launch alternatives. Despite being an idea we'd really, really like to be good, it's solving the wrong problem. You need speed, not altitude.

The aircraft-launched rocket is a bit more interesting, and has already been employed by Virgin Galactic and others to reach space. Basically, you get a gigantic plane (or, in one proposal, the equivalent of two 747s strapped together) and you tie your rocket to the bottom or top. You get up as fast and high as you can go and launch the rocket. The idea is that by starting out with some speed and altitude, you can save some propellant. Also, because you're launching from a plane, you don't have to worry so much about weather conditions. If the weather is bad, just fly the rocket to somewhere with smooth air.

The problem is that you're only getting a few percent of the speed you need and a few percent of the altitude you need. So the savings are likely to be very small. And the trade-off is that you have to launch a big, hot rocket from a megaplane, which is going to add to cost and complexity while limiting the size of the rocket. SpaceX, which has already shown its interest in changing the way space launch works, rejected the aircraft launch method for just these reasons.

Method 6: Space Elevators and Space Tethers

Imagine you've got a big rock spinning around Earth. Attached to the rock is a ribbonlike cable, about 62,000 miles long. It goes all the way down to the surface of the Earth, where specially designed elevators take cargo, travelers, and spacecraft up and down.

This idea may seem outlandish, but it's been quite well studied (particularly by former NIAC fellow Dr. Bradley Edwards), perhaps because it would represent the ultimate solution to space travel needs. All the problems with all the other methods are fixed by a space elevator, though not without adding some new and unique challenges.

Having vehicles going up and down the cables means you can send fuel as you go. It means you don't have to accelerate quickly. No discarded parts, no dangerous explosives, and no slamming into an unwelcoming atmosphere. You just ride to the orbit you want, picking up speed relative to Earth's surface because the tether itself is already in orbit.

Here's roughly what it might look like:

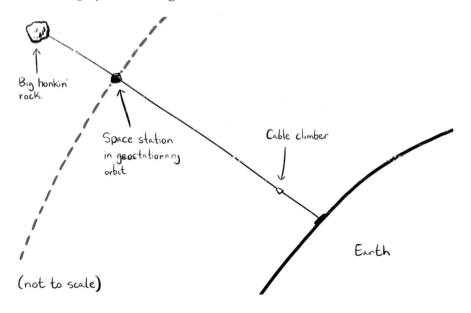

Big honkin' rock.

Space station in geostationary orbit.

Cable climber

Earth

(not to scale)

The main pieces are a counterweight, a cable, and a base station.

The counterweight is there to make sure that the center of mass for the whole system (including the enormous cable) is at what's called geostationary orbit.

If a thing is in geostationary orbit that just means that if you are sitting at the equator and looking up through a telescope, the thing will always be at the same point. It's revolving around the Earth as fast as the Earth is turning. And it's at a particular velocity and distance that means it naturally keeps whirling around Earth without needing a boost at any time.

We'll skip the orbital mechanics, but the basic results are that the cable is relatively taut, but not so taut that it rips apart; and the counterweight doesn't loop around the Earth, tying the cable around the equator like a giant spool of thread.

How you would get a counterweight is another matter, but there are three common proposals—capture a near-Earth asteroid, gather a lot of the space crap we've left up there over the years, or just have the cable be sooooooo long that its sheer mass will hold the cable taut. We find the idea of a gigantic asteroid base to be the most romantic, so we'll go with that.

Down from your asteroid base, you've got a cable that has a high specific strength. This means it's strong enough to resist breaking, but it's also very lightweight. This is important, because if the rope is strong but heavy, its own weight will pull it apart. If the rope is weak but light, it'll snap the first time it encounters rough conditions, like the high atmospheric winds of Earth.

Supposing you can build the cable, the last part is your base station down on Earth. Most proposals call for a moveable sea platform. This is because a moveable platform can maneuver away from bad weather and can adjust the position of the cable in order to avoid space junk higher up. Also, out at sea *there is no law.*

Well, okay, there's some law. In fact, it's called the Law of the Sea. But none of it pertains to cables going to space.

The laws that'll govern a space elevator are actually pretty important. It was our sense that most scientists who work on this stuff would like the space elevator, if it's ever built, to be something that no single nation controls. If one nation alone has cheap space access, that's a pretty big power asymmetry. So, from a let's not all kill each-other perspective, having joint ownership of the means of cheap launch might be good.

Once this system is operational, researchers estimate that cargo could go to space for under $250 per pound, very fast and very safe.

As a bonus, once you build one, building another is a lot cheaper. After all, the big initial expense is going to be launching all that cable by conventional means.

Most likely, we will also have base stations along the way up. These can serve as fuel and maintenance depots, as well as launch points for satellites and spacecraft. One of the best features of this design is that you can reach different altitudes just by climbing up and down the cable. Once you reach 300 miles up, you're in Low Earth Orbit, like most satellites. Go a lot higher and you get to geostationary orbit, which is great for communications satellites, but right now costs a fortune to reach. Beyond that, you get to where Earth has very little gravitational pull. So you're like a rock at the tip of a sling. If you want to get fired into space, just hop out of the station.

This last point is especially exciting for those of us who watched a lot of *Star Trek*. If you can get anywhere you want simply by climbing (instead of carrying fuel onboard), not only is satellite launch cheap, launching very large spacecraft is cheap too. More than any other method that seems feasible this century, the space elevator would open up the solar system to human exploration.

So why not do it?

Well, there are a lot of technical challenges, but the greatest of all is what the hell do we make the cable out of?

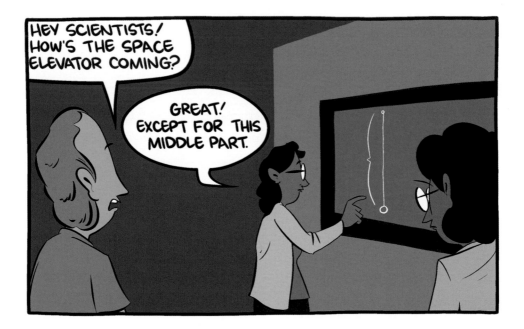

The unit of specific strength is the Yuri, named for Yuri Artsutanov, a pioneer of the space elevator concept, whose last name was apparently too hard to pronounce. Depending on who you ask, an ideal cable material should be 30 million to 80 million Yuris. For reference, titanium is about 300,000 Yuris and Kevlar is about 2.5 million Yuris. Regular materials will not do.

The most promising candidate material is called carbon nanotube. Imagine a molecule made entirely of carbon atoms, but shaped like a straw, with its width a small fraction of the thickness of a human hair.

It turns out that if you have pure carbon nanotubes with no imperfections,[11] they can get into the 50- or 60-million-Yuri range, meaning they *might* work as a space cable. The problem is that carbon nanotubes are a relatively recent discovery, and we're still pretty bad at making them. The longest nanotube ever created was made in 2013, and made headlines all over the place, and . . . it was about a foot and a half long.

You can, of course, weave these fibers together, but the smaller the pieces of the weave, the worse your specific strength becomes and the more imperfections there are likely to be. A long, taut cable is only as good as its weakest part, and if your cable breaks at any point, someone in a cable car is gonna have a real bad afternoon.

The long-term question is whether a market exists to make better and better materials. According to NIAC's Dr. Ron Turner, "Theoretically, and materials-wise, the carbon nanotube could become plenty strong enough . . . for a space elevator. Terrestrially, there wasn't much of a market after a certain point, so the carbon nanotube fibers have not continued to grow as strong as the space elevator would need."

Even supposing we could get the fibers long enough, Mr. Derleth points out an issue for carbon nanotubes: "The material is very sensitive to electricity, and so if it ends up having a lightning strike happen, it will disintegrate a large portion of the ribbon. . . . Thankfully, there's a solution to this; unfortunately, it's not a very satisfactory one, intellectually. There is an area of the Pacific Ocean

---

11. This "no imperfections" part is extremely important, as slight flaws in the carbon nanotubes can dramatically reduce the strength of the cable.

that has never had a recorded lightning strike. So you'd place your space elevator there. That's the solution. Now if a storm came through, you would have a lot of worry."

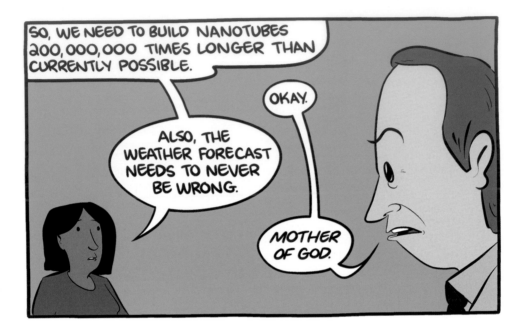

If you could keep cable rope away from lightning strikes, you would still need to worry about debris. There's a lot of stuff zooming through space, so even if you can dodge the big stuff, little things might wear out the cable over time. According to Dr. Turner, "This concept of continuing to have to refurbish the elevator remains one of its biggest challenges, in my mind, and it's one that they don't have a good answer for, yet."

Plus, the space elevator might make for a particularly good target for terrorists. Dr. Phil Plait (astronomer and author of the blog *Bad Astronomy*) points out that someone coming along to snip it might not be such a remote possibility. "It's a pretty ripe target for people to want to destroy, and not everybody is nice. We have enemies."

We're guessing a lot of you would like to know what happens if you have a cable to space, and then somebody comes along and snips it. Among the people we interviewed, there was some disagreement on how bad this might be. Dr. Turner and Mr. van Pelt thought that a break in the space elevator tether might not be so catastrophic. They point out that groups have tried modeling what would happen by simulating the results that follow snips at different points. Roughly speaking, it's something like this:

Anywhere you cut, the stuff above the cut will go into a higher orbit and the stuff below the cut will fall toward earth. The stuff in higher orbit will need to be collected, since it represents some serious space trash.

If you snip high, then a lot of the cable falls in toward Earth. Once that happens, there are a number of complex interactions between gravity, the atmosphere, the motion of the Earth, and possibly some electric charge picked up from the solar wind.[12]

---

12. "Solar wind" is the term for the stream of charged particles the sun fires in every direction.

The mechanics get a bit complex, but in short, the cable will start to whip-lash back and forth, heating up in the atmosphere, until it breaks apart. Because the material is necessarily lightweight, the individual pieces probably won't hurt anyone down on the surface. And you could minimize the risk even more if the cable were made into a mesh comprised of thinner strands.

Dr. Plait agrees with some of these particulars, but is a little less optimistic about the implications. "Sure, stuff hundreds of kilometers up might burn up as it falls (not that thousands or millions of tons of material burning up over one area is a great thing to have happen), but what about stuff from lower down? That'll just fall. And then there's the space debris. Most of the tower is below orbital speed, so it'll all fall down to Earth, but 35,000 kilometers of it will fall through the orbital space of Low Earth satellites. I have NOT done the math or physics here, but until someone can tell me how that won't destroy hundreds or thousands of assets in space I'm not inclined to think a space elevator is a great idea."

Concerns

Cheap access to space means our relationship to space will change forever. It will be possible to create large space stations or even settlements in orbit. We see this as a good thing, but it could potentially put power in the hands of bad actors. One idea that originated in the Cold War was the so-called rod from God. Basically, you get a heavy hunk of metal and throw it from space at an enemy. Given its weight, height, and whatever speed bump you can give it, a simple metal rod could do as much damage as a nuclear bomb. Right now, the only people who go to space are ultraqualified supernerds—the sort of people who pass psychological tests and are willing to spend decades training for a chance to get a few months in space. If space becomes more generally populated, we could be putting ourselves in a dangerous position.

Setting aside terrorists, another scary possibility is how we might deal with the ambitions of powerful nations. Outside of the Soviet breakup, the national borders on Earth have been relatively stable since the costliest war in human history ended in 1945. The laws of space that are legally agreed upon essentially say that no nation can claim anything out there. We find it hard to believe that a nation with a space elevator would abide by this. In fact, as we'll see in the next chapter, the United States is already making a few moves in this direction.

We tend to think of the universe as divided between space and "down here." But this is sort of like an ant thinking Earth consists of "space" and "inside the anthill." It's true, but perhaps a bit chauvinistic on the part of the ant. "Space," as we use it, refers to everything in the entire cosmos outside one planet in one solar system in one of many billions of galaxies.

If humanity gets cheap access to space, it's hard to imagine there will be no conflict over claims. And, as seems likely, if only one (or a few) nations have that access first, it may create conflicts on Earth. In other words, if humanity gets cheap space access, it means there may be a sudden political squabble at the same moment a single nation gains the most powerful weapon system in history.

Another concern is ecological. In the near term, going to space is probably going to involve incremental improvements to fuel-intensive methods like spaceplanes and rockets. Some of these fuels are relatively harmless, while others are particularly nasty polluters. According to Mr. van Pelt, the environmental damage "depends on the type of fuel. For instance, the Space Shuttle main engines ran on liquid oxygen and liquid hydrogen, with the resulting exhaust being superheated steam.[13] So in the end, it is just water coming out of those engines. But the Shuttle's solid rocket boosters (or any solid rocket booster) are another story, not nice indeed. And releasing water vapor at very high altitude, like the Shuttle did, can apparently also be harmful." This isn't a huge deal right now because we don't launch many rockets. But if reusable rockets make space launch cheap and common, they could be a serious environmental risk.

There's also the orbital environment to worry about. Since Sputnik, we've

---

13. Even if the only inputs are liquid oxygen and liquid hydrogen, it still takes a lot of energy to create, store, and transport them. If these things are accomplished by renewable power, or by the fusion reactor of our dreams, then the launch is indeed "clean." If you're using coal to get your propellant, there's still a pollution problem, even if it's not present on launch day.

thrown more and more stuff into space. It's starting to get crowded, with the rate of collisions increasing. Cheap access to space may mean more space debris. That said, if space launches get cheap, we might be able to invest in some sort of space cleanup vehicle.

According to Mr. van Pelt, "It becomes really an economic issue because if you've got your multi-hundreds-of-millions-of-dollars telecom satellite and it gets damaged by some debris, I mean there's a real price tag on that. You get insurance, the insurance goes up because there's more and more space debris."

In the longer term, cheap access to space would make space settlements more feasible, which might result in genetic differences between Earthbound and non-Earthbound humans. Mr. Derleth notes, "It turns out that the mathematics of genetics is different for small, isolated populations. Large populations can have more genetic mutations than small ones, but a small population can spread mutations to the group faster. So, you can imagine, if we had a thousand people on Mars and the colony was self-sufficient, well . . . it's very expensive to send more people, right? So there might not be too many new people coming, especially as a percentage of population. And the colonists would be having babies— true Martians—and the little tykes would be growing up at ⅓ G and with little atmosphere and even less of a planetary magnetic field to protect them from radiation. So it's possible the colonists would get a lot of genetic mutations, more rapidly, from the potential radiation exposure and growing up in ⅓ G, despite being a small population. . . . At some point there might be 'Martian humans' and 'Earth humans,' and society would have to try to interpret what it means to have two different kinds of humans."

I mean, technically we already feel this way about people who talk at the movies. But point taken.

How It Would Change the World

As we read books by people who'd been through the exciting era of Apollo, we got a sense of the frustration people felt as the hopes of a space-age future crashed into the economic reality of rocket launch. If we want those space-age dreams—massive, fast spacecraft with enormous crews, settlements all over the solar system, and trips to distant stars—we need to bring the cost way down.

Most of the books and papers we read about nonrocket spaceflight attempted to calculate the cost per launch their system could provide. The lowest we saw was around $5 to $10 per pound, with more conservative estimates more in the $250 to $500 per pound range. If even the conservative goal can be achieved, humanity's interactions with the rest of the universe would be forever changed.

In terms of commerce, here's one way to think about it. A typical space elevator proposal is for once-per-day climbers to be able to lift 40,000 pounds into orbit. The International Space Station weighs about 900,000 pounds. That means that even if the elevator operator takes weekends off, we could launch a huge space station once a month. And the cost would be something like $5 billion total, instead of the currently estimated final price tag of $100 billion.

Cheap launch would also facilitate big improvements in satellite systems, which should mean better communication methods and really, really accurate GPS systems.

It could also help with global climate change. Scientists have estimated that just a few percentage points more cloud cover might entirely offset all the warming expected to occur over the next century. One way to artificially do this would be to launch a large screen to block some of the incident light. One day you may look up to a friendly dark patch floating in the sky, protecting you from catastrophe. Ideally, it would say something like "FOR THE LOVE OF GOD, HUMANS! HOW COULD YOU LET THIS HAPPEN?!" on the Earth-facing side.

Speaking of humans, we might go up to space just for fun. Right now space tourism is so expensive and so controlled that (as far as we can tell) all the private space flight people are slightly crazy, very nerdy billionaires. We're happy they're doing their thing, but it might be nice if a mere millionaire could get in the game. The possibilities for space tourism are probably pretty sizable. The few space tourists who've been able to hitch a ride have spent around $20 million for the privilege. Not bad considering most people spend a lot of their time in zero G puking.

Yes, puking. As we discussed earlier, astronauts experience "weightlessness" because they're in free fall. Another time you experience free fall is when you start zooming downward in a roller coaster. Your caveman ancestors didn't have a lot of experience with satellites and roller coasters, so you aren't evolved to deal with it very well. Your stomach isn't used to food floating around in every direction, and your sense of balance isn't used to a world where you somersault every time you lean back. This is why the International Space Station keeps a ready supply of barf bags, even for trained space professionals.

In a space elevator, as long as you don't go too high, you're experiencing something pretty close to normal Earth gravity. This is true even if you're as high as a satellite like the International Space Station.

But wait, didn't we say people in the International Space Station don't "feel" any gravity, because they're in free fall? Why don't people in the elevator feel the same way?

The short answer is that in the elevator, you're going around the Earth much more slowly. You know the Earth rotates once every twenty-four hours, because that's how often the big bright thing in the sky comes up. It follows that your space cable also has to go around once every twenty-four hours—any faster or slower and it'd start wrapping around the Earth like a thread on a spindle. By contrast, the International Space Station is going around the Earth so fast that it sees a new purple-red sunset every ninety minutes. That's sixteen times the romance per day.

In the International Space Station, the ground, so to speak, keeps moving away from your feet as the space station curves around the Earth. Your space elevator doesn't go fast enough to pull off that trick. So as you rise up the elevator cable, you mostly lose the feeling of gravity simply by getting farther and farther from Earth. If you want to go above the atmosphere for a dramatic view of the stars above and the sky below, you can probably do it without a barf bag.

So when do you start puking? That's up to your stomach, but what we can tell you is when (and why) you experience total weightlessness. To experience weightlessness, you want to be in free fall. You want to be moving at such a speed that you keep "missing" the Earth even as you fall toward it. This gets easier to do as you get far away from Earth, because you've got a lot of room to fall into, and because the Earth doesn't yank nearly as hard on you. For any given distance from Earth, there's a particular speed you need to reach in order to loop around it in a circle.

With respect to Earth's equator, the space elevator never changes speed. It's always turning around Earth once every twenty-four hours. But, as you go farther up the cable, you eventually reach the height at which the elevator's speed of rotation matches the speed required for it to stay in free fall. This particular speed and distance is the geostationary orbit we referred to earlier, and it's very special. If you toss out a sign that reads EARTH IS FOR BUTTHEADS, not only will it orbit forever, but it will stay at the same position in the sky for any observers. It's too far to see any regular-sized objects with the naked eye, but any observer who points a telescope at the right point on a cool evening of stargazing will always see your douchey signage.

Perhaps the most exciting possibility of all will be the sense of adventure that would come with inexpensive spaceflight. Some people wonder why the dreams of the space age never became a new age of exploration. The core problem is that it's just too damn expensive, which means it's largely been a public-sector project. In the modern world, that means there's a pretty serious aversion to risk, as Rand Simberg argued in his book *Safe Is Not an Option: Overcoming the Futile Obsession with Getting Everyone Back Alive That Is Killing Our Expansion into Space.*[14]

As long as space travel is extremely expensive, and not merely *pretty darn expensive*, you're not going to get people willing to take wild risks and go on bizarre adventures. More to the point, even if there are (as we suspect)

---

14. Best subtitle in history.

astronauts willing to take a one-way trip to Mars,[15] such a program would never be approved.

All this is to say, we hope we haven't made you pessimistic about the possibility of fundamentally changing the way we get to space, and thereby fundamentally changing our relationship with the universe. It will not be easy to make any of these technologies work, but once they do work, we can finally turn heaven over to the adventurers.

Nota Bene on Gerald Bull and Project Babylon

Gerald Bull did not have an easy childhood. When he was still a boy, his mother died. As the Great Depression struck Canada, his father remarried and sent his many children to live with different parts of the family. Bull was lucky to end up with family members who were well off enough to get him into university at a young age. He studied aeronautics and quickly gained a reputation as a brilliant engineer and as someone who would do anything to finish a job and finish it cheap.

In the 1950s, Canada was trying to develop a domestic missile program called Velvet Glove. The project had a great deal of trouble attracting talent, as students went to the United States for higher wages and greater prestige. Bull, at this time a strident Canadian patriot, was willing to stay behind and stand by his country. And so, still in his twenties, and looking young for his age, Bull was a major part of the Canadian missile program.

---

15. Also, there are a lot of regular people who are probably nuts. A project called Mars One, to send people on a one-way trip to Mars as a reality show, got over four thousand applicants.

But the Canadians were not big on funding their missile ambitions, so Dr. Bull had to get it done on the cheap. When he couldn't even get access to a wind tunnel he'd helped build, a friend suggested skipping the wind tunnel and just firing projectiles. The young engineer duly acquired an old 6-inch field gun and refitted it to fire missiles at 4500 miles per hour.

This is how Dr. Gerry Bull fell into ballistics, which led him to wonder if you could create a gun big enough to fire projectiles into space.

He was an extremely clever engineer, but had a reputation for hating to work with lesser minds—especially bureaucrats. According to *Wilderness of Mirrors* by Dale Grant, in the late 1960s, he once stormed out of a meeting with a Canadian defense minister, shouting that the minister had "the technical competence of a baboon."

Dr. Bull developed his supergun method as an alternative to rockets, but as he made more and more enemies in Canada, he had a harder and harder time getting funding. However, he had developed a few true believers in the American military and was able to leverage those connections into access to things America has in abundance: funding and surplus giant guns. With help from the United States' Department of Defense and Canada's (somewhat reluctant) Department of National Defence, Bull began what eventually became called Project HARP—the High Altitude Research Project.

With the kind of speed Gerald Bull was known for, the guns got bigger and the projectile designs got better and better. By 1962, from a massive gun installed in Barbados, Bull's team were making regular shots that went high enough to probe the upper atmosphere. Atmospheric data and research on high-velocity projectiles became a good source of funding for the project, but Dr. Bull had bigger goals. He believed that with the right modifications, a larger gun could fire satellites directly into orbit.

By 1965, they could fire significant payloads at about 7,000 miles per hour. This is a great start, but you need to hit 25,000 out of the barrel if you want to have a chance to reach orbit. Dr. Bull's idea was to "mate" a rocket to the payload, so the gun could get you most of the speed, while the rocket gave you a final boost to insert you into the proper orbit.

Things were going reasonably well when funding was pulled. HARP lost American funding as NASA pushed the U.S. Army out of space operations.[16] Then HARP lost Canadian funding due to the burgeoning peace movement, which didn't look kindly on a giant megagun.

Dr. Bull responded by forming a private space research company. It wasn't building mega-space-guns, but it was making money through a variety of government contracts. But on the side, Dr. Bull began working out ideas for a cannon to dwarf his old designs. A skyscraper-sized gun, it was to have a 64-inch barrel and be 800 feet long, able to fire 6-ton projectiles into space.

As the 1970s progressed, aerospace spending began to dry up. Dr. Bull, who

---

16. Given that HARP was a descendant of artillery guns, army funding was natural at first. As space became considered its own tactical domain, HARP became less and less appropriate as an army project. And, in any case, by this time NASA had already settled on rockets as the best way to go.

was perhaps less brilliant at business management than engineering, expanded the company more and more, which required greater and greater bank loans. In his struggle to keep things afloat, he entered the international arms trade.

Things soon unraveled. Dr. Bull got into a tangled situation involving the CIA, the Canadian government, and an illegal shipment of weapon parts and technology to apartheid South Africa. During a shipment through Antigua, the shippers informed the locals (who are mostly of African heritage) where the goods were going. News agencies got word, and it became an international incident.

A round of spy games for the sake of political ass-covering began, and Dr. Bull apparently was unable or unwilling to understand what was happening. He was brought up on charges of illegally transferring munitions. In a plea bargain, Dr. Bull's company was fined and he was sentenced to a year in jail, of which he served only four months due to good behavior. He fell into rage, depression, and alcoholism, feeling he'd been betrayed and scapegoated. As he served his time in humiliation, his company finally collapsed. All the holdings, all the technology he'd built, were sold off cheap. Basically, if you were trying to create a supervillain, this would be a pretty good way to go.

Dr. Bull, who had formerly considered himself a patriot of Canada, decided he was willing to work with anyone who would bring some version of HARP back.

Things get a little murkier at this point, due to the secrecy and disinformation of many parties, but in the late 1980s, Dr. Bull shows up in Saddam Hussein's Iraq, working on a supercannon called Project Babylon.

Let's say that once more: Dr. Bull shows up *in Saddam Hussein's Iraq, working on a supercannon called Project Babylon.*

This happened. Like, in real life.

The really weird thing is that the design for Project Babylon really does not appear to have had military applications. It was to be laid astride a mountain, which meant it could only fire in one direction. As it happens, that direction contained no significant enemy targets. It was also pointing in the proper direction (as the Earth spins) to fire into orbit.

In his book, *Bull's Eye: The Life and Times of Supergun Inventor Gerald Bull*, James Adams claimed that the evidence is clear that Bull was designing military weapons for Iraq in addition to the supergun. Given Bull's need for money, and his well-known ability to improve weapons systems on the cheap, the arrangement makes a twisted sort of sense. Why Hussein was willing to fund the supergun is harder to pin down. Saddam Hussein was said to have a messianic notion of his place in the Arab world, so the supergun concept may have served the dual purpose of giving Iraq the military utility of a cheap satellite launcher and the political capital that would go to the only nation in the region with a serious space program.

Or maybe they had the greatest miscommunication in history.

Western nations (and countries near Iraq) grew concerned. A slightly insane ballistics genius with access to a network of shadowy munitions companies was shipping tons and tons of metal parts and liquid propellant to a belligerent dictatorship. What could go wrong?

Apparently, somebody didn't want to find out.

In March 1990, Gerald Bull was found dead in a Brussels hotel, with $20,000 on his body. Given the sheer number of people who stood to gain from Dr. Bull's death, the potential enemies list is enormous. His own son suspected the CIA or Mossad, and according to *Arms and the Man: Dr. Gerald Bull, Iraq, and the Supergun* by William Lowther, "One CIA official, speaking on the strict understanding that he not be named, says it is the general understanding of Western intelligence agencies that Mossad gave the order to kill Bull."

There has been precious little written about Dr. Bull since the early nineties. In all likelihood, we'll never know for certain who cut short his strange and tragic career.

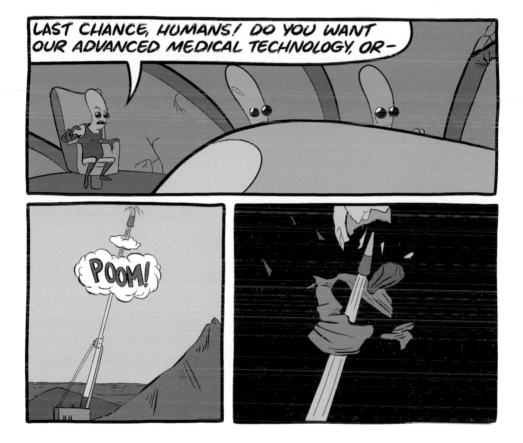

3.

Asteroid Mining

Rummaging Through the Solar System's Junkyard

The Earth was once a whole lot hotter. Long story short, that's why you can't have a house made of gold.

See, when you have a big, hot melty ball (like primordial Earth) in space, gravity tends to shift the heavy elements (like gold and platinum) toward the core, while sending the lighter elements (like carbon, silicon, and various gases) to the surface. The process isn't perfect, which is part of why you can find seams of heavy metal and metal ore near the surface. But by and large, the really fun stuff is hard to get. And the more we dig up, the harder it gets to find more.

This is where asteroid mining starts to look interesting. Asteroids are basically the junk that goes into making a planet, but they never permanently coalesced into giant space balls. This means they either never underwent the heating process when all the fun metals go to the core, or they *did* go through that process only to be blown apart later. So, just beyond Mars, there's this giant pile of planet rubble, with vast riches of metal and other resources that we might like back on Earth, or might use to build settlements in space.

If only we could find some slightly crazy people, with heads like engineers and hearts like pioneers.

Daniel Faber runs a company called Deep Space Industries. He has been a spacecraft engineer, the director and president of the Canadian Space Society, and has set up broadband in *Antarctica*.

According to Mr. Faber, there are enormous resources in asteroids just waiting to be mined: "There are asteroids that are made completely of metal, like natural stainless steel, nickel, and iron and . . . the smallest one we know in a near-earth orbit is 2 kilometers across. It goes by the glorious name of 3554 Amun and it contains in it more than thirty times the amount of metal that humanity has ever mined on earth. And that's one. And then, there are thousands of those. That's the smallest one that is in a near-earth orbit."

There have been very few missions to asteroids, so most of our data about them comes from telescopes and from examination of meteorites. Based on these observations, scientists separate asteroids into three main categories: C-type (carbonaceous), S-type (stony), and M-type (metal).

Carbonaceous asteroids contain a lot of stuff that's handy for human health, like carbon and water. Some C-type asteroids may be as much as 20% water, in some form or another. Water isn't going to be a huge seller back on Earth until hipsters decide they want asteroid cocktails, but it will be quite handy if you intend to set up a crewed settlement.

Stony asteroids contain, you know, stone. Specifically, they are loaded with silicates. Silicon isn't going to make you a fortune back on Earth either (Earth's crust is about 28% silicon), but it might be very useful for your mining settlement. Silicates have all sorts of applications, from glass to solar panels to a growth medium for plants, sometimes called "dirt." It's also called "earth," but that's gonna be considered chauvinist once we're living on asteroids.

Metal asteroids are mostly iron and nickel. Metals like these are very good for building in space, since, unlike rocky materials, they can bend and stretch without breaking. Metal is also nice for armor, and for printing money with your face on it after you declare independence from the damnable tyrants back on the mother planet.

In short, you have all the stuff for a successful settlement—oxygen to breathe, dirt to grow plants, metal to build, and water to make water balloons. Let's talk about financing.

As discussed in the previous chapter, it currently costs about $10,000 per pound to send something to space. That Big Gulp you're planning to take with you will run you about $20K. The Apollo 11 spacecraft, which only went as far as the moon, weighed over 100,000 pounds. So, not counting all the engineering, permitting, and staff required to build and operate a spaceship, you're starting this venture at roughly negative $1 billion. That's not great, but remember, buying an Airbus A380 costs about $400 million, so we're not entirely outside the realm of possibility.

One way to get asteroid mining to be *a thing* is to make some profit back home. For example, if you can mine platinum in space, you can sell it for about $18,000 per pound right now on Earth. And getting up into space is the expensive part. Getting back down is comparatively easy.

One of the (few) convenient things about open space is that, once you get away from heavy objects like planets and stars, you can get almost anywhere for cheap. Think about it: There are two main reasons it's expensive to fly from Los Angeles to Japan. (1) You have to climb over 30,000 feet against gravity and then fight gravity for the whole trip, and (2) there's a bunch of air slowing you down en route. Once you're in space, both these problems go away.

Thus, if you're planning to ship stuff back to Earth, the ideal place for a space base is somewhere with high resources and low gravity. Consider, for example, Phobos, one of the moons of Mars. Phobos is very small, so its escape velocity is a mere 25 miles per hour. This means you could set up a ramp on Phobos, drive a motorcycle up it, and fly off into space.

Earth's own moon has an escape velocity that's about 200 times greater than Phobos's. The result is that, in energy terms, it costs less to send a package from Phobos to Earth than from *Earth's* moon to Earth.

Asteroids are even better. A typical big asteroid has an escape velocity of about half a mile per hour. This means that if you succeed in creating an asteroid mining base, you can pitch refined asteroid contents back to Earth at very low cost.

That said, from our research and interviews, it seems unlikely that mining space for profit back home is going to work in the long term. Most of the stuff in asteroids is just not valuable enough outside of space. Although there's a lot of metal up there, it's not at all clear that it'd be cheaper to harvest asteroids than to create a similarly extravagant harvesting method down here on Earth. Plus, with the latter method, you could still hit the Burger King every day.

Potentially we could just harvest the rare, expensive metals, but this too is difficult. To the extent that there are rare, expensive metals in asteroids, you'd probably want to refine them in space to avoid having to transport and land a huge pile of ore back on Earth. But refining in space is tough because you need to build a refinery in space, and most current refining methods require gravity. You can simulate gravity by just spinning your refinery around really, really fast, but that'd require a lot of energy to start, and you'd want a pretty huge setup so you don't feel like you're on a carnival ride 24/7.

And even if you *did* find a giant asteroid made of gold, diamonds, and Mickey Mantle rookie cards, it's not entirely clear what would happen when you brought it all home. Physics behaves. Economics doesn't. According to economist Dr. Bryan Caplan of George Mason University, "Competition is the key. If a single firm hits the mother lode of platinum, it can hold back most of its stockpile to avoid flooding the market. If many firms share the mother lode, in contrast, each is likely to rush to sell off its platinum before its rivals do. Given the high fixed cost of space travel, asteroid mining is very likely to start with few firms, giving first movers a great opportunity to profit from whatever resources they find. Over time, however, success breeds imitation, so later generations of asteroid miners should beware."

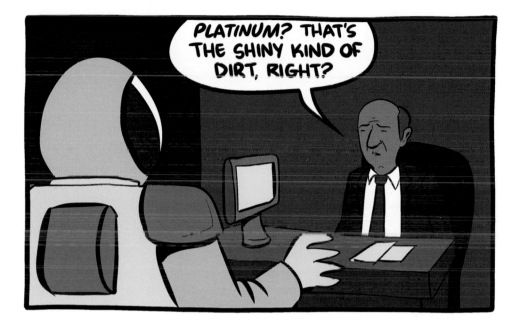

So, with all this difficulty, why are we talking about asteroid mining? Well, there's something bigger at stake here. It might not be worth it to ship a giant cargo of iron 280 million miles just so you can make new pipes for your toilet. But if you ever want to build a settlement in space that's more than a barely inhabited little bubble, you'll need resources. In space, dirt will be worth a lot more to you than diamonds and platinum, and it'll be a lot cheaper to lasso some space rocks than to blast dirt up from Earth.

As many scientists and engineers believe, moving humans beyond Earth is intrinsically valuable. There are mysteries and wonders out there. There are answers to questions we don't even know how to ask yet. There may be life like us or life unlike us, and it's hard to say which of those possibilities would be more staggering.

Pioneering the stars is a big dream, but the most successful pioneers are practical people. If you want to reach beyond this small planet, it will be cheaper, easier, and faster to use materials that are already outside of Earth's gravitational pull. The path out of the solar system may run through its junkyard.

Where Are We Now?

There are a lot of problems facing a would-be asteroid miner, so we're going to briefly touch on a few of the biggest ones.

You're going to need energy, both to get to the asteroids and to power your settlement. The biggest constraint is that, in the short term, you can't go to the hardware store. You need something pretty darn permanent. There are a lot of options, but probably your best bets among current technology are solar panels or a nuclear reactor.

Solar panels are nice because they can keep functioning for over twenty years. The biggest downside to them is that they don't produce a lot of energy for their size. But assuming you can get them up there, they're reliable, durable, and simple. It's even conceivable that you could repair and replace them on the asteroid settlement, since asteroids are relatively rich in silicates. Materials from asteroids can also be used to create mirrors to focus light on panels, increasing their energy output.

Nuclear reactors are good because they generate a lot of energy for their size, and they can last as long as a hundred years, assuming nothing goes horrifyingly wrong. Asteroids don't contain anywhere near enough uranium or plutonium to make a new reactor, but you don't need a lot of nuclear fuel to begin with.

A lot of research has gone into compact nuclear reactors, in part because they were at one point used on Soviet satellites. But, well, here's the thing about that: Suppose you launch a nuclear-powered ship into space and things don't quite go right. Suppose, for instance, the ship starts moving around dangerously and you lose control over it. Suppose it then breaks up in the atmosphere and dumps the remains of a nuclear reactor over a 400-mile-long swath of Canada.

Yeah, that happened. Cosmos 954, in 1977. So your nuclear reactor plan is nice and all, but you'll probably end up sticking with solar for political reasons.

Speaking of radiation, the sun is constantly firing off fast ions. While you're on Earth, you're protected by a thick blanket of air. You also have the Earth's magnetosphere, which takes a lot of the dangerous particles flying from the sun and either deflects them away from Earth, or routes them to Earth's poles so they can put on a little light show for people.[1]

The moment you leave home, you're getting bombarded by radiation.

The crewed missions to the moon were reasonably safe,[2] since the trips were very short, and the risks from radiation increase with the amount of exposure. But the asteroid mission will take a lot longer. Apollo 11 took a little over three

---

1. AKA aurora borealis and aurora australis.
2. Well, maybe they were safe. One study by Dr. Michael Delp at Florida State University found that Apollo astronauts (those involved in the missions to the moon) had a higher risk of death from cardiovascular diseases. Mouse studies suggested this was because radiation damages the cells that line blood vessels. However, given the small number of dead Apollo spacefarers, this study was only able to consider seven humans. Given what we've read about those early Apollo guys, there's a decent chance the heart disease had more to do with what we call the sex-booze-meat triangle than with what happened during a week or two in space.

days to get to the moon. Ceres, the largest object in the asteroid belt, is about 1,100 times farther away. Assuming we go just as fast as Apollo 11 (which, as lunar missions go, was *quite* fast), it'll take about ten years. That said, for longer-distance missions, you can probably get to a higher speed. For instance, NASA's uncrewed Dawn mission made the trip to the asteroid belt in about four years. Even at a relatively good clip, it's not exactly a day trip.

Plus, and this is a really cute feature of space travel, now and then the sun fires off megadoses of radiation, in what are called solar flares. You'll want to have a plan for that.

You have some options. You could armor the entire ship. A 2-inch-thick, 10-square-foot-wide piece of lead armor weighs roughly 1000 pounds. So, if you're launching everything from Earth, armoring your whole ship will roughly double the already enormous cost to get it to space. Another option is to use water as a shield. Water is actually an excellent radiation blocker, and you could have it do double duty by acting as a shield while it's stowed away for drinking. Of course, this raises the ominous possibility of someone eating too many pretzels, and then *drinking the radiation shield.*

Some scientists have suggested a "panic room." The basic notion here is that you're willing to deal with *some* radiation, but you want protection from those big flares. So you spend most of your time in the lightly armored body of the ship, but whenever there's a radiation blast, everyone runs to a small armored chamber. On the one hand, it'd probably work, and it'd cost a lot less than armoring the whole ship. On the other hand, you're now in a one-cubic-meter chamber with several other people, in the middle of space, and all of them are seriously reconsidering their life choices. Also, if someone needs to pee, remember there is no gravity.

You might say, "Well, I'll take some potassium iodide pills to protect myself from radiation." This is dumb. Potassium iodide pills only protect your thyroid. So later, when you arrive at your target, you'll have fourteen types of cancer and an arm growing out of your face, but sure, your thyroid will be in tip-top condition.

But let's suppose you've got a radiation solution. Your next problem is landing on an asteroid. This is easier said than done. There's a very good chance that the asteroid is spinning. If the asteroid is big enough, you can try to find the axis and land there, but this is a pretty delicate maneuver. Another option is to stop the asteroid from spinning before landing.[3] This is tough for two reasons.

First, it costs energy to slow down a spinning object. The bigger it is and the faster it spins, the more energy you're going to need. There isn't any real way around this; it's fundamental physics. You either change the spin or you find a way to deal with it.

Second, you may have mediocre information on its internal structure. Many asteroids are "rubble piles," meaning they are essentially clumps of rocks and dust held together by a little gravity and by intermolecular forces. This makes them hard to stop from spinning. If you try to stop the spin, you may just blow away a chunk of the asteroid. Given the instability of rubble pile asteroids, this could result in embarrassing outcomes, like horrible death in a cold silent void.

This leads us to your next problem—how do you land safely? Remember, there is pretty much no gravity on an asteroid. If you try to touch down on it, you might just bounce back off. Scientists have a few ways to get around this, such as drilling, firing a harpoon, gluing, clasping, using geckolike sticky "feet," and more. What's tough is that no solution works well under every circumstance. If the asteroid has a deep layer of dust, drilling won't get you a good hold. Gecko feet won't help much either. If the surface is hard metal, the harpoon will bounce right off and probably eject you into space. If the surface is flat and rocky, your claspers will just scrape it and eject you into space. Ideally you know what you're up against, but unless there has been a lot of reconnaissance,

---

3. That said, at least a few scientists think the speed and rotation of some asteroids could be a resource itself. For example, Dr. Alexander Bolonkin proposes a large net that grips a moving asteroid, steals some of its energy, detaches, then flies onward. Think of it like a skateboarder grabbing onto a rotating carousel in order to get some speed.

you don't. This is why more recent missions to asteroids carry multiple landing technologies.

One of our favorite ideas for a lander was proposed by Dr. Karen Daniels at North Carolina State University. She thinks a lander might work like the roots of a plant. "Anybody who weeds a garden knows how hard it is to pull a plant directly out of the ground, roots and all."

The basic idea is that you have little diggers that work their way in between the bits of rubble. As they successfully dig into the surface, they can also link up for added support. This system sounds entirely workable to us, but it is disappointingly un-Freudian, compared to the harpoons and drills of more obvious methods.

One proposal to get around the landing problem is to just net the entire asteroid. Like, in a net. A huge giant space net. This isn't as crazy as it sounds—remember, we're in microgravity so it doesn't take a lot of force to move things around. If you have a very strong material, you can potentially fold the net into a very small space, then unravel it when you get to the asteroid. Once the asteroid is netted, you can use the net as a landing surface for your settlement, or use it to haul the asteroid to some other destination. One such project confirms our suspicion that space people spend 90% of their time creating acronyms. It is called the Weightless Rendezvous And Net Grapple to Limit Excess Rotation (WRANGLER) System (proposed by Tethers Unlimited's Dr. Robert Hoyt).

Another proposal made by TransAstra Corporation is called APIS (for Asteroid Provided In-Situ Supplies). APIS captures an asteroid in a bag, and then uses concentrated sunlight to heat and cut away at the asteroid. They call this process "optical mining." This method causes a release of the water from inside the asteroid, which you might want after you accidentally drink the radiation shield for your space base.

Okay, but say you have all the technical stuff worked out. You've still got a major problem: It's not clear who has mining rights for those space rocks. Sure, there's the 1967 Outer Space Treaty, but that treaty doesn't deal with private ownership rights. It *does* say sovereign nations can't claim anything in outer space. It's not clear if there are any laws governing your space settlement, as long as you're don't give in to your temptation to declare independence from Those Damnable Earthlings.

In November 2015, the U.S. Congress passed H.R.2262—U.S. Commercial Space Launch Competitiveness Act. The act stipulates that "A U.S. citizen engaged in commercial recovery of an asteroid resource or a space resource shall be entitled to any asteroid resource or space resource obtained, including to possess, own, transport, use, and sell it according to applicable law, including U.S. international obligations." In other words, "America can't claim space, but Americans can." Indeed, Congress was careful to note in the act that America is not asserting ownership over any celestial bodies. If and when Americans start mining space rocks, we'll see how other countries feel.

So you see there are a lot of difficulties. On the plus side, all these difficulties are really, really awesome.

In terms of an actual mission, there have been several uncrewed asteroid visits by NASA, ESA, CNSA and JAXA, the American, European, Chinese, and Japanese space agencies respectively. A Japanese craft called Hayabusa managed to collect a tiny amount of asteroid dust and return with it in 2010. The successor craft, Hayabusa 2, should bring back more sometime around 2020. A somewhat similar NASA mission called OSIRIS-REx should return to Earth around 2023.

There are a number of proposed missions, including ones at NASA, to capture a whole asteroid or even to send human beings to a near-Earth asteroid. So far, none of these more extravagant plans has received the funding to make it happen.

Concerns

A major concern at the moment is figuring out the particulars of law and order in space. At some point, these asteroids will need to be policed. You'll essentially have giant resource pools floating around in space, and once the technology comes along to make capturing these asteroids and extracting their resources easier, no doubt we'll end up with space-crime by space-criminals. As cool as this sounds, you might not feel great about it if *you're* the one with a space-knife in your space-back.

The amazingly named Dr. Elvis (he has a first name, but . . . *come on*) of the Harvard-Smithsonian Center for Astrophysics says, "Eventually, we'll need a space geek-squad to repair billion-dollar mining equipment in space. I reckon we'll also need space sheriffs and space forensics experts too, because rare, valuable resources always lead to 'extra-legal' activities."

Dr. Elvis also points out that we'll be spoiling environments that have been untouched since the dawn of our solar system. This is a general problem with all space exploration, but in this case we're also explicitly planning to obliterate the object of interest.

Perhaps, suggests Dr. Elvis, *whose name is Dr. Elvis,* in the future we'll come up with a system of asteroid parks to preserve at least a subset of them, but this issue will have to be tackled legally too. There are some things that only form in space and aren't found on our planet. Much like with modern rain forests, the risk we run is not only of destroying things we want, but of destroying things we've never even seen.

One other concern is safety. Even if humanity can agree to allow private space mining, what are the rules on where you can move asteroids?

If asteroid-moving technology were widely available, it could put a dangerous weapon into the hands of undesirable people. We suspect this would be a relatively low risk, given that even history's craziest tyrants still seem to care about their own lives. Even assuming you could perfectly drop an asteroid on Washington, D.C., the results for the planet would be hard to predict. It's possible a big strike would kick up enough dust to blot out the sun, cool the Earth, and destroy a year of crops. But the fact that "Hey, Kim Jong-un isn't *that* nuts" is pretty cold comfort.

The largest encounter between Earth and an extraterrestrial object in recorded history was the 1908 "Tunguska event," when a huge object (we don't know if it was a comet made mostly of ice, or an asteroid made mostly of rock) exploded over rural Russia. This object was probably about 120 feet in diameter, and its explosive yield would've been comparable to 185 times the bomb that was dropped on Hiroshima.

The Tunguska meteorite is small compared to some of the asteroids we would be able to bring home.

How It Would Change the World

I believe the only way to explore the solar system at the scale it demands is to have a space program that grows without limit. That is what economies do. Let's harness capitalism to bring us out to meet the worlds around us.

• Dr. Martin Elvis

If it turns out that you really can turn a profit on Earth from mining in space and returning the products, it would turn deep-space travel from something that only governments can afford to something comparable to the trucking industry. The economic benefits of cheap materials would be incredible, and the ability to inexpensively travel to space would be better still.

That said, based on our research, we think this is unlikely. A better bet for changing life for all of humanity is simply this—by using the resources found in asteroids, humans can begin to settle space, and we can dramatically increase the pace at which we explore our solar system.

Mr. Faber pointed out that the resources in space are vastly greater than those found on Earth: "The materials we can access on Earth in the deepest mines are 3 to 4 kilometers deep. We can get to oil that is maybe down to 6 or 7 kilometers. But if we take the materials that we can reach on all the continents in the world and bunch it into a sphere, it would be about 200 kilometers across. That is all the material that we're ever going to have to work with on the surface of the earth. In space, there is hundreds or thousands of times as much material. So not only can we support hundreds or thousands of times as many people off that material, but also it's readily accessible. It's floating around in free space. We don't have to dig deep into dangerous places to get it, we just go and effectively disassemble the asteroids and send them to the kinds of places that we want to live."

If we have all this material available up in space and can figure out how to manufacture things in space, then the cost of space travel goes down dramatically. Gigantic space colonies become a feasible option. And asteroid mining could help us get around once we're already in space. Water and carbon collected from the asteroids can be turned into rocket fuel, which means we can use resources collected in space to help us get these resources back home, move from colony to colony, or explore farther out into space.

This is what really excites Mr. Faber: "I eventually got bored with racing solar cars, and hang gliding and windsurfing and such, and decided that the biggest benefit I could create for humanity, the most important thing that was going to happen in my lifetime was moving humanity off Earth and becoming a multiplanetary species or an interplanetary species."

We, uh . . . we're also totally bored with racing and windsurfing, which is why we are sitting in this office editing this document . . . for *humanity*.

SECTION 2

Stuff, Soonish

4.

Fusion Power

It Powers the Sun, and That's Nice,
but Can It Run My Toaster?

Nuclear fusion is the ultimate solution for human energy needs. It's clean, it can use common elements as its fuel, and it carries no risk of catastrophic meltdowns. Yet today, the reason your toaster successfully makes Pop-Tarts is probably that someone nearby is burning coal or natural gas.

Before we get into the complexities of how you're destroying the world every time you make a toaster pastry, let's talk about the physics of fusion for a minute. Fusion is when two atoms fuse. That's it. In the context of fusion power, we're usually talking about two atoms of hydrogen fusing.

Hydrogen is the lightest element in the periodic table, and its nucleus is just one charged particle, called a proton. But as with all elements, there isn't just one form of hydrogen. There are many, and the different types are called isotopes. What's different about different isotopes? The number of chargeless particles in their nucleus. These are called neutrons.

About 99.98% of all hydrogen is in a form that has zero neutrons. Technically, this is called hydrogen-1. Or, for fancy kids, you can call it protium. But in practice we usually just call it hydrogen. About 0.02% of hydrogen has an added neutron in addition to its proton. This stuff is generally called deuterium, because the Greek word for "the second thingy" is *deuteros*. If you have one more neutron beyond that, you've got tritium, which pretty much doesn't exist anywhere because those extra neutrons make it unstable. Tritium has a half-life of 12.32 years, meaning that if you have a jar full of tritium, when you check back in 12.32 years, half of it will be gone.

It's sort of like how the most common form of relationship is two people, but trios and higher numbers exist. So if you're at a fusion power conference and you get invited back to someone's hotel room for "a rare isotope," your answer is "yes."

You can go higher up, to hydrogen-4, hydrogen-5, and so on, but all these substances are extremely unstable, lasting a tiny fraction of a second on the rare occasions they pop into existence.

Why do we care about all these isotopes? Because bigger hydrogen isotopes have a much easier time fusing. To understand why, imagine you need to ram two cars into each other, but each car has a giant magnet on the front. The magnets are set up so that the left car and right car have the same magnetic pole at their front. So as they get close, they really strongly resist each other.

However, if you get them extremely close, each car has a small latch that holds them together so powerfully, that the repulsive force of the magnets can't pull them apart. Now ask yourself: If you wanna get this right on the first try, would you rather have the cars be Mini Coopers or SUVs?

Intuitively, you want the heavier car, right? You know from experience that slowing down a heavy car is a lot harder than slowing down a light one. Assuming the magnets are the same strength, if your goal is to successfully latch the two cars together, you want the cars to be as heavy as possible, so the magnets will have the hardest time stopping them.

Hydrogen atoms face a similar situation when you try to fuse them, which we'll get to in a moment. In short, they repel each other quite strongly until they get close. By adding more neutrons, you get hydrogen that's more SUV than Mini Cooper.

When hydrogen isotopes fuse, they become a different element—helium.

This may seem weird, but it's no weirder than how two pieces of bread make a new thing called a sandwich. Hydrogen is the element with one proton. Helium is the element with two protons.

For our purposes, the interesting thing is this: When these isotopes transmute from hydrogen to helium, they release a HUGE amount of energy.

Here's why: The atomic configuration we know as "helium" needs less energy to hold together than two of the atomic configuration we know as "hydrogen." When fusion happens, that energy has to go somewhere. When it does, we can harvest it.

This idea that the energy "has to go somewhere" may seem a little odd at first, but it's really nothing too strange. To see why, let's talk about a crossbow. If your crossbow's string is loose, you know that it's different from when your crossbow's string is taut. The two crossbows (loose and taut) are essentially the same, but due to their physical arrangement, the loose string is pretty much useless, while the taut string can kill that one guy you thought was the black knight when you were drunk at the Ren Faire.

In order to get the string taut, you have to use energy. And in order to loosen the string, you have to release energy.

"Is there really more energy in the taut crossbow?" you ask. YES. Yes, there is. In fact, suppose you took the loose crossbow and the taut crossbow and put each separately in equally sized vats of acid. When each was finished dissolving, the vat that dissolved the taut crossbow would be just a little bit hotter than the vat that dissolved the loose crossbow.

Yeah. Remember when you were bored in physics class? That's because your teacher wasn't dissolving crossbows.

So, when you think about it, getting energy by changing the configuration of something isn't all that weird. You encounter situations like this all the time. A rock held high can break someone's foot, while a rock on the ground can't. A stretched spring will spontaneously change shape when you let go of it, while a loose spring won't. Two magnets forced to touch north pole to north pole will jump apart when released, while two magnets with opposite poles touching will just sit there.

All this is to say that getting energy by bringing two atoms together is not entirely foreign to your everyday experience. The way you configure all the parts of a system determines the future of that system. And sometimes, you can set up that system so it does things humans like.

As it happens, the change in configuration from two small atoms (hydrogen) to one larger atom (helium) gives you an energy release. For most fusion reactions, you use at least one extra-neutron hydrogen isotope. Generally speaking, the released energy comes in the form of one of those extra neutrons getting kicked away at high speed after the isotopes combine. Once you have hot, zooming neutrons, you can capture that energy in the same way a regular old steam turbine works: Smack the neutrons into water, heat it into rising steam, and turn turbines.

But wait. Why do we need fusion to turn a turbine? You can do that with coal or diesel or gas or wind. That's true, but fusion power is special in that it takes only a tiny amount of fuel, and that fuel is relatively abundant.

According to Garry McCracken's book *Fusion: The Energy of the Universe*, the lithium[1] in a laptop battery and the deuterium in "half a bath of water" could generate 200,000 kilowatt-hours of fusion energy, equivalent to 40 *tons* of coal.

So why don't we have fusion power yet? Even though it's easier to get big isotopes to fuse relative to their smaller nonisotope counterparts, it's still really hard. This is because of something called the Coulomb barrier. The Coulomb barrier is the huge amount of energy (at least, on the atomic scale) required to get two protons really, really, really close to each other (think of the magnets in our earlier car analogy). Protons repel each other. In fact, as they get closer, the force of repulsion becomes greater. Sort of like when you try to get two socially awkward friends to talk at a party.

But when protons get really, really, really close together, another force called the strong nuclear force comes into effect. This force is very strong over short distances. It overcomes the repulsion, latching the protons together (like the latch in the car analogy).

In other words, getting protons to fuse is like getting your two socially awkward friends to marry each other. If you can just get them next to each other, they can share their contention that Advanced Dungeons & Dragons 2nd Edition is the best edition.[2] They'll be so in love, they'll never part. But until they are right next to each other, they won't even make eye contact.

---

1. Lithium, the third element on the periodic table, is split to create tritium in a lab setting.
2. It is.

HOW NERDS AND PROTONS WORK

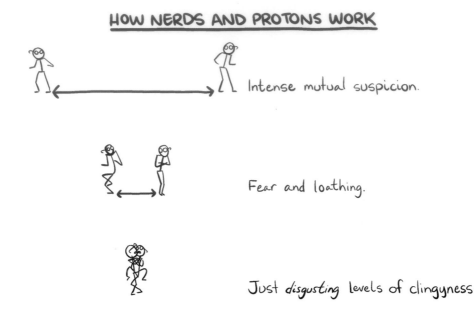

Intense mutual suspicion.

Fear and loathing.

Just *disgusting* levels of clingyness

The sun is powered by fusion. So why can't we do whatever the sun does, but in a lab here on Earth? Well, the sun has an advantage: extremely high gravitational pressure at its center, which means that hydrogen atoms are constantly smacking into each other at high speed. We're not sure of the implications for our nerd-dating analogy,[3] but the point is that these conditions don't exist on Earth.

To get fusion energy, we have to figure out some way to get the force of all that gravity via some means *other* than gravity.

HOLD ON. WAIT. Don't we already know how to make fusion bombs? Isn't that what they used in that Bruce Willis documentary about blowing up an asteroid? Why can't we do this:

---

3. Okay, wait: imagine Joss Whedon just got dropped into the middle of Comic-Con with nothing to defend himself but a box of spoilers for the next *Star Wars* movie.

1. Take hydrogen bomb.
2. Explode.
3. Collect heat.
4. Boil water.
5. Spin turbines.
6. Power toaster.
7. Eat Pop-Tarts.

The answer is . . . actually, yes, we could do this. We can create sunlike conditions with a fusion bomb. Essentially what you do is detonate a bunch of good
ol'-fashioned fission-based[4] atomic bombs, and then use the resulting energy to
compress a sphere filled with hydrogen. Thus, your hydrogen is smacking into
itself really fast, not because of gravity, but because of all the force from the
atomic bombs. A bunch of fusion events occur, and BOOM—enough energy to
punch a hole in the atmosphere or turn a big patch of desert into glass.

But there are some problems. First, the container for the bomb would probably not survive the blast, making it hard to capture that energy. Second, even
if the container did survive, it'd be incredibly irradiated. Third, if the container
were imperfect, you might send a massive cloud of radioactive dust into the atmosphere.[5] Fourth, the Russian Embassy is calling, and they sound a little irri
tated on the phone.

---

4. Fission is when an atom splits apart. When large atoms, like those of uranium or plutonium, split, energy
is released. In the right setup, the splitting of one such atom can cause several others to split, and their splitting causes yet more splits, and so on. This chain reaction can be used to generate power or to create bombs.
5. See this chapter's nota bene.

Geopolitics and safety concerns rule out the simple approach.[6] We have to turn elsewhere for sunlike conditions. Like, for instance, to that spherical metal cage on your tabletop.

Okay, well, maybe not *your* tabletop, but definitely the tabletop of Richard Hull, the first amateur scientist to achieve fusion. Mr. Hull coruns a Web site called Fusor.net, where people can learn how to create and run tabletop fusors.

While the site has been joined by lots of "fusioneers," only about seventy-five have officially demonstrated fusion and thus earned the honor of joining the Neutron Club. A boy named Taylor Wilson joined the Neutron Club at age fourteen. So, you know, keep that in mind when you're trying to decide if you've accomplished enough in your life.

If you know what you're doing, and have access to all the right equipment, you can make one of these yourself for about $3,000. That's right, kids. For just $10 a day for 300 days, *you could have a neutron gun.* Here's how:

---

6. Arguably, sheer expense rules it out too. And you would probably prefer for there to be a lot of bureaucracy and paperwork between an energy company and a nuclear warhead.

First, go to the spherical metal cage store and grab a spherical metal cage. Now, electrify that cage so it has a very strong positive charge.

Inside the positively charged cage, put a smaller cage that has a very strong negative charge.

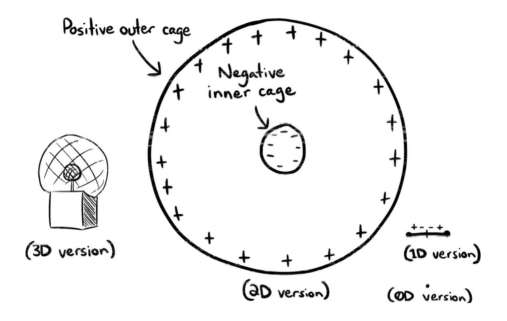

Positive outer cage

Negative inner cage

(3D version)

(2D version)

(1D version)

(0D version)

Having done all this, you put the whole thing in a vacuum chamber, then add some deuterium gas between the two cages. Purchasing the deuterium gas may require filling out some paperwork for Homeland Security, but if you don't want to get put on a list, just go online and buy some "heavy water."

Okay, so that sounds like the worst scam in the world, but we promise it's real. Water is two parts hydrogen and one part oxygen (H_2O). Heavy water is just the same, only the hydrogens are replaced by deuterium (D_2O). Remember, deuterium has an extra neutron, so the water, despite looking like regular old water, is a bit heavier.

Now, take your D_2O and run electric current through it. This will separate it into oxygen gas and deuterium gas. At this point, peek out your window to make sure the FBI isn't on the street. All clear? Good. Here comes the hard part—you have to "dry out" the gas by removing the heavy water vapor. This, Mr. Hull tells us, is quite difficult, since even a small amount of vapor will ruin your fusion party. There are different ways to do this. For example, you can pass the gas through a cold tube, where the vapor is more likely to stick to the cold walls, or you can just run your gas through something with a lot of surface area to stick to, like cotton.

Now, take your deuterium gas and inject it into the electrified cage. The powerful electric field separates the gas into positive and negative parts. Now you have little deuterium nuclei, *ripe for smashing*.

The positive outer cage pushes the deuterium toward the center, and the negative inner cage pulls it toward the center. Suddenly, a whole lot of collisions are happening.

Remember our socially awkward nerds earlier? Imagine a room. In the center of the room are rare, early *Star Wars* action figures (specifically, the figures with the telescoping light sabers that were discontinued in 1978). Around the walls of the room are posters of scenes from *The Hobbit* movies that *weren't in the original book*. Now, between these horrible, *totally-not-canon* posters, and these awesome, *super-rare* action figures, you throw a pile of nerds.

The nerds all run from the wall posters and toward the action figures. The combined attraction and repulsion is so strong that they blow *right by* the toys, only to collide with each other and embrace, at which point they reflexively discuss lost episodes of *Doctor Who,* and thus they convert from your awkwardly single friends to your uncomfortably clingy friends.

Just so, the deuterium particles between the positive (repelling) outer cage and the negative (attracting) inner cage lock onto each other and fuse. Hooray.[7]

But wait . . . earlier, didn't we say fusion was really hard?

Yeah, so here's the thing. Most of your socially awkward friends aren't getting married. Sorry. This couple over here disagrees about who knows more *Wolverine* trivia. That couple over there simply clonked their heads together and passed out. And the guy in the yellow cardboard armor is "saving himself" for Sigourney Weaver.

Just so, most of the deuterium particles don't successfully fuse. Some hit the cage. Some miss each other. Some hit each other hard enough to deflect, but not hard enough to combine.

Because you threw enough of them together, you'll get *some* fusion, but you're only benefiting from the law of averages. Your tabletop fusion device is net negative on energy. It takes in more energy than it puts out.

---

7. Dr. Alex Wellerstein, whom we talk about later in this chapter, suggested having a baby fly out as a metaphor for the neutron. This is, of course, ridiculous.

You may be wondering about the fusioneers at this point: Why are they doing fusion if they aren't going to get net energy out of it? Turns out fusioneers tend to come in three types: those who think they are going to solve the fusion problem and give the world clean, cheap energy, those who are doing fusion as the coolest DIY project in all of history, and those who want neutrons for research. Incidentally, Richard Hull is in the latter group, spending his retirement probing the nature of the atom, as one does.

And now we've arrived at the nub of the problem: We can make fusion happen by using a bomb, but it's really hard to control. We can make fusion happen with a simple electrified cage, but it's not very efficient. And right this second, we don't have a great way to split the difference.

There are several different approaches to solving this problem, but in the most mainstream experiments, there are two ways to go about it: You can either try to blast all your fusion fuel at once, or you can confine it into a small place as it heats.

Blasting, probably with a laser, is good because you get a lot of fusion events to happen all at once. When individual fusion events happen at different times, this tends to slow down the overall fusion reaction. To understand why, think about our nerd couples. If every couple who falls in love clasps hands and runs out happily, they'll knock away incoming nerds and these fallen nerds won't have a chance to fuse with other nerds.

Just the same, fused atoms make tiny little energy releases, which mess up your fusion reaction by pushing against incoming protons. So it's good to try to get all of your fusion reactions to happen at once.

The downside to the blast approach is that you need an incredible amount of energy to be released in a small amount of time. Like, more power than the rest of the world put together, for a small fraction of a second. This isn't entirely crazy—there's a device at Sandia Labs, called the Z machine, which stores a huge amount of energy in a massive capacitor bank, then releases it all at once. But we'll get to that in a moment.

The second approach—confining and heating—has the advantage that the fusion fuel can be heated over a longer period of time in a more controlled fashion. To handle the fact that you're not blasting it all at once, you need to confine the fuel really well.

So you use plasma. If you don't remember plasma, it's a type of matter you can achieve at high temperatures.

Think about it like this: You start with an ice cube. You heat it until the molecules are no longer attached to each other, but they are still attracted by their intermolecular forces. In other words, you go from solid to liquid. If you keep heating, eventually the energy in those water molecules gets so huge, they don't care about the mild attraction between molecules and just fly off in every direction. That is, they become a gas. If you keep heating way, way, way, way hotter, the atoms themselves get ripped into two parts—their positively charged nuclei and their negatively charged electrons.

For this discussion, ignore the electrons and imagine a plasma as a very hot gas that has a positive electric charge. This charge makes plasma very different from gas. If there's a bad smell in your room in gas form, you can only get rid of it by blowing air against it. If there's a bad plasma, you can control its motion with magnetic fields. Not only can you get rid of it, you can shape it. You can confine it.

There's this one minor downside for "magnetic confinement"-type reactors, which is that it can be a little tricky to confine plasma that's roughly the temperature of the inside of the sun.

With these constraints in mind, there are a number of experiments, large and small, attempting to find a solution.

Where Are We Now?

There are too many experiments to go through all of them in detail. So, we're going to confine ourselves to a few.

NIF

The National Ignition Facility (NIF) is so awesome it was used as a set for *Star Trek*. You could probably *literally* get nerds to fuse inside it.

NIF is working on a technique called inertial confinement fusion. The way they do it is this: Start with a superpowerful mega-ultra-laser—a laser so powerful that melting its own lenses is a problem. That laser is split into 192 beams.[8] The beams converge on a gold cylinder about the size of your fingertip. Inside this cylinder is your fusion fuel. The cylinder absorbs the massive dose of energy, then emits extremely powerful X-rays into its center, compressing the fusion fuel and (they hope) causing a lot of fusion events. It's like a cute, little bitty hydrogen bomb.

---

8. You have to hit the fusion fuel from multiple sides all at once to make fusion happen. So why use one giant laser split 192 ways instead of 192 little independent beams? The reason is timing. It's much easier to synchronize 192 beam parts than to fire 192 beams all at once.

NIF has not yet achieved "ignition," which is when you've produced more energy than was required to start the reaction. As of 2015, they've hit about a third of that goal.[9] But, it's not called the National One-Third-of-Ignition Facility. A great deal more research is needed to validate this fusion method.

MagLIF

Another exciting experiment is the MagLIF project at Sandia Labs. MagLIF is short for Magnetized Liner Inertial Fusion.

Here's a rough sketch of how it works: Take a chilled cylinder filled with fusion fuel. Now, take a powerful laser and blast the fusion fuel through one side, so it reaches a high temperature very quickly. Before the fuel can blow out through the cylinder, use your incredibly gigantic capacitor bank to cause an

---

9. The actual number is arguable. NIF defines "power in" as the amount of power that goes right into the gold cylinder, *not* as the amount of power required to fire the shot. By that standard, they're closer to 1/100 of ignition.

enormous electrical discharge. This creates a magnetic field, which collapses the cylinder.

In short: Blast it then smash it.

Initial experiments with this approach have already been conducted and have shown a great deal of promise. However, it's not yet anywhere near breakeven (which is the point at which the energy you put in to start the reaction is equal to the energy you get out of it). I mean, it's doing better than the fusion experiment in our basement, which is made of magnets, gum, and hope. . . . *But* this project is really exciting because the lab has produced some detailed computer models suggesting that if the approach is scaled up enough (in terms of energy), it should be able to achieve breakeven. Experiments are ongoing, and their giant capacitor bank (the Z-machine we mentioned earlier) is being expanded. If things go well, it's possible they'll make it by the end of the decade.

ITER

The biggest experiment of all, using the most successful and well-studied fusion configuration, is ITER, the International Thermonuclear Experimental Reactor.

ITER uses a "tokamak" configuration, which is a Russian acronym for toroidal chamber for magnetic confinement. Basically, imagine you have a gigantic Dunkin' Donut, only instead of sawdust and tears, the inside is filled with plasma. Using magnetic fields, you confine the plasma so it flows around the middle of the tube. Imagine a thin ring of plasma inside the giant donut. Due to the strong magnetic fields, it is extremely hard for the plasma ring to escape. With the plasma confined, ITER uses several methods to heat it: They electrocute it, microwave it, and fire a beam of neutrons at it. As more energy gets transferred into this tight band of plasma, the plasma gets hotter and hotter and hotter, until (if it works!) the plasma protons start fusing.

Here's the neat part: The fusion reaction of the plasma releases energy, which causes more plasma to heat, which gets you yet more fusion. Think of it like a candle. If you hold a lighter to a candle, but the wick doesn't catch fire, all the heat in the wick will go away once you turn off the lighter. But if you can get the wick to catch fire, the candle will keep burning until the entire thing is gone. The fire creates more fire. In the same way, if you get the fusion reaction going fast enough, it can sustain a continuous fusion burn. If that happens, all the external heating methods could be shut off, and the reaction would continue.

And bonus, it's happening inside a huge metal donut.

ITER is the biggest, most expensive fusion project today. Unfortunately, ITER (like many science megaprojects) has been fraught with delays and cost overruns. You know how it's hard to get cats with brain damage to agree on something? Well, imagine if instead of brain-damaged cats, it's political appointees from many different nations. The current cost estimate for ITER is over \$15 billion,[10] which is a bit of an increase over the initially projected \$5 billion. In fairness, they were only one digit off.

Still, progress continues. As we write this, the actual tokamak portion of ITER is finally being built. The hope is that the full-on fusion reactor experiment will happen in 2027, just in time for the first Robot Uprising. ITER is generally considered the best hope for a functional fusion reactor any time soon, and not without good reason. The biggest currently running tokamak, the Joint European Torus (JET), is already hitting 60%–70% of breakeven.

Other Projects

In addition to these big experiments, there are many smaller heterodox approaches. The scientists we've talked to have generally responded to these experiments by saying some version of "I hope it works, but it probably won't." But just for fun, we're going to tell you about an interesting project by a company called General Fusion.

Their method wins the award for being most Mad Sciencey. It works like this: Take a sphere and fill it with heavy liquid metal. Spin the sphere so fast

---

10. We've actually seen lots of different projections, some ranging as high as \$50 billion.

that a vortex opens in its center. Inject fusion fuel into the vortex, then have a series of rams simultaneously bash the outer surface of the sphere. The rams create a pressure wave that moves through the liquid metal toward the center, where it collapses the fusion fuel.

If it works, one nice feature is that energy can be harvested from the heated liquid metal.

A number of companies are working on similarly offbeat approaches, but many of them are kept secret enough that we can't really get interesting details about them. For instance, Lockheed Martin recently announced that they also have a fusion reactor that is quite close to working. But they've kept the details pretty vague. This is the scientific equivalent of saying, "Hey, I have this really sexy girlfriend you've never met, but I don't have any pictures and you're not allowed to read our e-mail." It could be true, but we're not getting excited just yet.

Concerns

Any fusion reactor design will produce some radioactive waste. When people hear the word "radioactive," they imagine a laughing corporate executive dumping neon green ooze onto a day care, then turning into a swarm of bats and flying away into the night. This characterization is not, in its particulars, entirely accurate. ITER won't produce any liquid waste like a fission plant would, and the radioactive tritium will be captured and reused for future reactions.

In fact, the ITER Web site states that things are so well contained that a fire in the tritium plant wouldn't even justify evacuating the local population. Yes, parts of ITER that are bombarded by the reactions will get irradiated, but these parts won't qualify as being "highly" radioactive and will cease being radioactive over a relatively short timescale.

Nevertheless, for those who are concerned, it is important to note that a nuclear *fusion* reactor is really nothing like a nuclear *fission* reactor, which is what people are referring to when they talk about nuclear energy. Both involve atomic nuclei, but so does your mom, Saturn, and apple pie.

According to Dr. Daniel Brunner at the Plasma Science and Fusion Center at the Massachusetts Institute of Technology (MIT), "The fusion reaction only produces helium and neutrons, no greenhouse gases and no long-lived radioactive waste." Neutrons stop when they bump into the side of the reactor, and helium is a nonreactive gas. Plus, kids will be really happy when their birthday balloons are held aloft thanks to thermonuclear reactions. According to Dr. Brunner, "Unlike fission waste, this material is safe enough that within about a hundred years it can be recycled."

Another frequent concern about fusion is whether there is a risk of meltdown. Simply put, there is none. In case it isn't already clear: It's really, really, really, *really* hard to get fusion to work. At least from the perspective of people not wanting to get blowed up, this is a positive feature.

Dr. Bruce Lipschultz of the University of York told us, "Only enough fuel for the immediate reaction is kept in the reactor at any given time, unlike a fission reactor, where a year's worth of fuel is stored within the reactor. And a fusion reaction is such a challenge to maintain that if containment is lost in the reactor . . . the fusion reaction will be blown out like a candle in the wind." Since he works on tokamaks, it's more like a giant plasma-filled donut in the wind, but we take his point.

None of the people we interviewed brought up concerns in terms of environmental or social problems. The real problem with fusion seems to be simply getting it to work. In fact, Mr. Hull is fond of saying that "Fusion is the energy of the future, and it always will be."

We talked to Dr. Alex Wellerstein, a nuclear historian[11] at Stevens Institute of Technology, about this. He was slightly more optimistic about the technology itself, but had concerns about its market potential. He pointed out that even with existing technologies, we need specialized carriers for the fusion fuel, sometimes made of expensive metals like gold and beryllium. The fusion fuel itself might be expensive, depending on what technologies are successful. Thus, even if we're getting more energy out than we put in, we have to consider the cost of the energy, and how long it'll take to return on its investment. Even if the payoff isn't clear in the short term, research may still be warranted. For example, it took about seventy years for solar photovoltaic cells to go from a lab creation to a practical way to build a power plant.

As Dr. Wellerstein says, "We need to break out of just funding things that will give immediate returns or else the long term will never get here."

With the development of major technologies, a big stumbling block is always funding a risky venture. At the height of the Apollo missions, the U.S. government spent close to 4.5% of its entire budget on NASA, employing over 400,000 people. Without that huge outlay of funds, it's possible that today we'd be saying that moon landings are always fifty years away. By comparison, the domestic fusion energy budget for the year 2015 is roughly the same

---

11. Meaning he studies nuclear power, not that he has nuclear powers.

amount it costs to buy a single 747 Jumbo Jet.[12]

As with any technology, it's possible there are serious ramifications that have not yet been well considered, but at least right this second, a working and economically viable fusion reactor seems to be a pure good for humanity. Here's hoping that last sentence doesn't show up in a postapocalyptic history book about human hubris.

How It Would Change the World

Simply put, fusion energy is an essentially limitless source of fuel. No greenhouse gases, no long-lived radioactive waste, and no chance of meltdown.

· Dr. Bruce Lipschultz

If fusion energy could be made economically viable, the most obvious benefit would be cheap energy. Although a fusion reactor would require maintenance and upkeep like any other power plant, given the abundance and relatively low cost of the input, it should be an energy source whose price lowers over time as technology improves.

Because energy is an input in literally every product, we might expect a lower

---

12. This doesn't include U.S. funding for international collaborations. The U.S. government also contributes to ITER, but as you might expect, some American politicians aren't thrilled about sending money for an over-budget science project being built in France. The U.S. continues to make yearly contributions on the order of $100 million to $200 million dollars, but representatives have repeatedly threatened to block this, as costs have escalated. Even American scientists are a bit concerned about large contributions to ITER, because the large foreign expenditure is crowding out funding for domestic facilities.

price for most consumer goods. This would be especially true for industries that are very energy intensive, like the manufacturing of chemicals, cement, paper goods, and metals.

The environmental benefits would also be substantial. Fusion reactors would produce minimal pollution and no carbon, which would mean we could get energy without contributing to climate change. Even if fusion were more expensive at first, when you factor in the potential environmental damage of ever greater atmospheric carbon, it might still be the better route economically. Also, the most likely fuel for fusion is hydrogen, which is the most abundant matter in the universe. So acquiring fuel should have a relatively low environmental impact. In other words, if you had your own fusion plant, you could burn perfectly good solar panels in your fireplace for fun, and *still* be invited to Greenpeace meetings.[13]

Fusion energy could also be a great help to long-distance space travel. Due to safety and efficiency considerations, most spaceships use liquid fuels and solar panels for energy and propulsion. An onboard fusion reactor would be relatively safe in the case of an accident, and would provide a large amount of energy. In addition, hydrogen is found throughout the solar system, and even in the relatively empty regions beyond. If a spaceship could collect this material, efficiently separate out the deuterium, and fuse it, that ship would have a potentially unlimited service life.

There are probably also a lot of hard-to-predict benefits of cheap energy. It would've been hard to imagine gasoline-powered cars by the millions back in the age of whale oil. Just the same, in this age of relatively expensive energy, we may not be able to see what wonders are over the horizon.

---

13. Not guaranteed. In fact, Greenpeace is on record as being opposed to ITER, though not for the reason you might think. They feel that we should be spending the money on modern renewables, like solar and wind power.

Nota Bene on Project Plowshare

It's southeastern New Mexico, 1961. Dry, flat land with scattered scrubby plants. A place that has seen few humans over the past millennium is, this day, populated by United Nations representatives invited by the White House. They are here to witness an atomic explosion. Although it may have geopolitical implications, the point of this bomb is not war—it is to demonstrate that atomic bombs can be used for peaceful purposes, that the atomic sword can be beaten into a plowshare.

Here was the idea: Drop a bomb down a deep shaft of rock salt way out in the desert. Detonate. The salt will reach extremely high temperatures and become molten. You now have a huge volume of molten salt, which is an excellent way to retain heat. In theory, you can use that heat to make steam, drive turbines, and run your toaster.

As in previous explosions under what came to be called Project Plowshare, this one produced an impressive shock wave and shook the dust from the desert ground. Then things got weird. When nuclear bombs are going off, *you don't want things to get weird.*

As scientists, reporters, military members, and government officials watched, a white vapor began to seep up from the blast site. A deadly cloud hung in the wind as all the assembled visitors were ordered to get to their cars and leave the scene.

What went wrong? The explosion was stronger than the scientists expected, and the resulting gas managed to break through to the surface. How radioactive was the gas? We don't know. Why?

Gophers.

Nope, that's not a typo.

GOPHERS.

Once more.

GOPHERS.

Yes indeed, a couple of friendly burrowing rodents (whose ultimate fate can probably be guessed) chewed through electric cables at the blast site, which took some of the radiation detectors off-line.

As far as your authors know, this is the only known gopher-related nuclear accident, but it is nevertheless illustrative of why Project Plowshare never quite got off the ground. In theory, it was a beautiful idea: There are all sorts of uses to which massive, cheap explosions could be put—creating harbors, digging a new Panama Canal, and generating new elements.

If this sounds crazy, that's for two reasons: (1) It's crazy, and (2) this was the age of nuclear optimism! Walt Disney released "Our Friend the Atom," in which Tinker Bell makes the atom symbol with her magic wand. Ford released the Nucleon—a concept car design for a future nuclear vehicle.[14] The jury was still out on just how dangerous atmospheric radiation was.

Thoughts like these led to what came to be called Project Plowshare. Between 1961 and 1973, under this program the United States conducted thirty-five individual nuclear detonations as part of twenty-seven tests. Each explosion had multiple research objectives, from examining products of the reaction to just straight up seeing how big a boom we could make.

It turns out we can make quite a boom. A number of the project craters can still be visited to this day. However, the greater legacy left behind by these blasts was one of resentment by local people, many of them indigenous Americans, who were treated callously. The scientists and bureaucrats in charge of Project Plowshare repeatedly ignored local concerns, underestimated the damage they would cause, and overestimated their ability to deliver a cheap nuclear blast.

The silver lining to the nuclear cloud was that it sped the development of the modern environmental movement. One of the strange things about this project was that they repeatedly hired environmental scientists to determine whether the blast would cause problems. These scientists found that, yes indeed, ecosystems and the people who live in them don't like giant radiation doses. At this point, invariably, the people in charge of Plowshare brushed them off. This re-

---

14. Imagine a road trip where you go twenty years before refueling!

sentment among intelligent professionals fused with resentment among targeted communities to produce some of the first modern environmental protests.

Like so many nuclear projects, even up to modern times, the history of Plowshare is one of a potentially great technology's developments being set back by a bunch of smart people acting like morons. The level of callousness was impressive. According to one story, lead scientist Dr. Edward Teller once suggested (as a joke, but still) using a sequence of nuclear explosions to blast a polar bear–shaped harbor into Alaska. There was, in fact, a serious proposal to bomb a harbor into Alaska even though it wasn't particularly desired by the locals, given the potential dangers and the fact that it was frozen over most of the year anyway.

Although we aren't recommending atom bombs for any construction projects, the later results are somewhat tantalizing. Before the final test in 1973, the technology actually worked relatively well. For example, one of the later parts of the operation, called Project Rulison, was to test to see if a fusion bomb could liberate natural gas. Guess what? If you have a bomb that's twice as powerful as the Nagasaki blast, you can access a lot of natural gas. The local radiation level afterward was surprisingly low—only about 1% higher than it had been previously. This was a potentially huge result. And there was hope at the time that the bombs used for peaceful purposes could be made relatively "clean." Hydrogen bombs use both fission and fusion reactions, but the vast majority of the nasty by-products come from the fission part. One of the major components of Plowshare was finding ways to mitigate those by-products.

Despite some promising results, by 1975 Plowshare was shelved. The scientists and engineers had made a lot of progress, but too many things stood in their way. The nascent environmental movement opposed them, and given the bureaucracy required to get a nuclear weapon to a test site, atomic bombs weren't saving much, if any, money compared to conventional bombs. And in any case, they hadn't produced much of use. Their biggest development was liberated natural gas, but companies weren't too keen on trying to market gas that's only a *teensy bit* more radioactive than average.

Moreover, in what has to be one of the more ironic turns in all of history, the invention of hydraulic fracturing ("fracking") midcentury undercut the idea of using atomic power to release natural gas. Tell this to your environmentalist friend if you want to break her brain. Adding to the irony, the ever-more-expensive war in Vietnam made funding less available to nuclear tests for fun and profit. So yeah, you may have been spared from a more radioactive world thanks to . . . fracking and Vietnam.

Over the course of the 1950s, '60s, and '70s, the danger posed by radiation became better and better understood, and people become more concerned not just about the quantity of radiation, but also the type. One of the most important findings was the danger of a major by-product of nuclear blasts— strontium-90 (Sr-90). You can think of Sr-90 as sort of like calcium, only it emits radiation. Because it gets absorbed into bone like calcium, Sr-90 is a particularly dangerous radiation source. Especially ominous was the discovery of Sr-90 in baby teeth, during the years of frequent nuclear testing.

Yep, baby teeth. Drs. Louise and Eric Reiss did a brilliant piece of research, in which, over a twelve-year period, they collected lost teeth from schoolchildren, eventually amassing about 320,000 of them. Teeth are mostly calcium, so lost teeth are a pretty easy way to check for how much strontium we're absorbing into our bodies. They found that around the peak of nuclear testing in 1963, when atmospheric strontium was quite high, kids had about fifty times the normal level of strontium in their teeth.

This finding partly motivated the Limited Test Ban Treaty (LTBT) between the United States and the USSR. This treaty probably posed the greatest obstacle to Project Plowshare, because one of the LTBT's rules was that you could detonate bombs in your country, but the effects of the bomb couldn't *leave* the country. This may not sound restrictive, but in practice it can be. If I set off a firecracker, some of the resultant carbon monoxide will eventually spread over Russia. The same is true of radioactive dust. Although neither the United States nor the USSR took the treaties as seriously as they should have, it added more bureaucratic complexity to the already (reasonably) complex task of getting a nuclear bomb to see if you could build a harbor.

For instance, the treaty stipulated that bombs could not be detonated underwater. Later, Plowshare people wanted to detonate a bomb under the seafloor, as opposed to in the water. This led to a real debate over whether "underwater" literally means "under water." Such is the glamorous life of foreign diplomats.

As Soviet–U.S. relations improved and it became clear that hydrogen bombs did not represent a great cost savings, Americans decided that atomic weapons would be better left on the shelf for use in case of apocalypse. Thus ended Project Plowshare.

In case you're wondering, the Soviet Union had a comparable program, called Nuclear Explosions for the National Economy, which lasted right up until that nation's dissolution. If you go to Google and search for "Lake Chagan," you will find an oddly circular lake in Kazakhstan, which was formerly a Soviet republic. You can probably also find a video that shows the instant of its creation. And, creepily enough, there are some propaganda shots of a guy swimming in the lake shortly after it was filled from a nearby reservoir. Here's hoping he had good health insurance.

5.

Programmable Matter

What If All of Your Stuff Could Be Any of Your Stuff?

Question: Why do you use your computer so much more than your bike? Answer: Because you're a creepy shut-in.

Okay, but, the other issue is that your bike more or less does only one thing: it moves forward when you pedal it. Your computer does many things, and (more important) it is *capable* of doing infinite things. That is because your computer is already "programmable matter." It can run any program, display any image, make any sound, connect to any device (well, if you can find the cord and Windows didn't break it). And these programs, sounds, images, and so on are not permanently embedded into the computer, like with a photograph or a record or the physical mechanism of a steam engine.

This is why a person from the year 1900 would find most of your stuff to be familiar. Your broom is made out of plastics, but it behaves the same as his wooden one. Your washing machine is a clever contraption, but nothing about it is hard to fathom. Your computer? Thaaaat's where he decides you're a witch.

What if we could make all of your stuff like the computer? Or, at least, why, in this age of powerful computing and advanced synthetic materials, can we not make a lot of your stuff adapt to your desires? Why can't the materials in a building automatically respond to changes in the weather? Why can't you order four chairs to reshape themselves into a table? And for God's sake, why do we have to fold origami ourselves instead of just yelling at the paper until it turns into a crane? These things may not be as far off as they sound.

MIT's Dr. Erik Demaine explains his enthusiasm about programmable matter like this: "To me the exciting thing about programmable matter is the idea of making gadgets that could serve many functions. I can imagine my bicycle turning into a chair when I wanted to sit down and not ride around. Then it becomes my laptop. Or my cell phone unfolds into a laptop. . . . We live in a computational world where software is reprogrammable. . . . Programmable matter represents doing the same thing for hardware. . . . If you want to get the latest cell phone, you have to go out and buy some physical stuff. In the future we can imagine that the same stuff we have can rearrange itself into a new model. That's the dream."

Scientists, engineers, and artists from around the world are attempting to make some version of this possible. Some want to program matter in the sense of designing it to respond to conditions around it. Others want robots built into everything we do. At their most ambitious, these creators dream of globs of matter that can morph into just about any shape. As computers and electromechanical systems become ever smaller and more efficient, there may come a day when you can reach into a pool of shimmering liquid and pull out any device, from a wrench to a phone to a robot pet.

Why do we want this? Well, there are practical reasons (which we'll get to) but on some level we suspect human beings just really love things that change into other things. Like, consider the *Transformers* series. It's about life forms from distant planets with superhuman minds and alien biology, fighting a war near planet Earth. And why do we find this exciting? Because in addition to all that, they can turn into *really sweet cars*.

Where Are We Now?

Programmable matter is about endowing physical objects with information. This may be literal bits and bytes in an onboard computer or it may be "knowledge" embedded in the structure of an object, via its shape and material makeup. This makes the field of programmable matter pretty diverse.

We're going to start by talking about materials that are "programmed" in the sense that they engage in complex specific behaviors based on how they were constructed. For example, exposure to water may cause one of these programmed materials to bend in a predetermined way, even though they don't have any onboard sensors or computational ability. Then we're going to talk about origami robots and reconfigurable houses, before we move toward more exotic attempts at creating universal robots—basically the T-1000 from *Terminator,* only ideally they're not trying to kill us all.

Programmed Materials

Professor Skylar Tibbits at MIT is the guy for programmed materials, and here is how he envisions the field: "The idea from my perspective is you want to program matter, program physical entities to transform themselves to change shape, change property, assemble themselves without human or machine intervention. . . . You want to somehow allow materials to transform themselves and often that's triggered by something in the environment—temperature, moisture, electroactive, some other trigger to allow them to transform." Professor Tibbits refers to this flavor of programmable matter as "4D printing" because you 3D print an object that changes through time depending on its materials and surroundings.

For example, there's the reconfigurable straw. The way it works is that you 3D print a straw, which has specially designed joints that bend when they take in water. By selecting how each joint will bend, in principle you can make just about any shape. Professor Tibbits made one straw that shapes itself into the letters "MIT" when put in water.

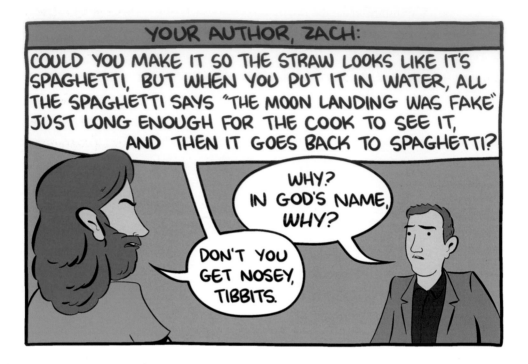

Another cool project that responds to the environment is the "HygroScope," created by the University of Stuttgart's Dr. Achim Menges and Dr. Steffen Reichert. The HygroScope is made of thin pieces of wood that bend in programmed ways in response to humidity. It is made up of hundreds of tiny pores that open and close as the wood bends in response to ambient conditions. Taken as a whole, it looks startlingly like some sort of enormous piece of alien biology, despite having no motors or computers built in.

Professor Tibbits sees two major hurdles to adopting these sorts of programmable materials: First, software is designed to work either with static structures or with robotic mechanical structures. He wants software that can design "materials that transform and are unstable and bend, curl, twist based on different activation energies."

The second problem is finding ways to make people actually want to buy this stuff. We hadn't really thought much about it before, but we kind of like that all the objects in the bathroom are "static." But Professor Tibbits believes there are ways that smart materials can be elegantly added to our everyday lives. For example, he was working on pipes that can pinch or expand in response to the amount of water flowing through them at the moment, to help control flow and respond to demand. Currently, he is collaborating with companies in the fields of sportswear, medical devices, packaging, and aerospace to bring programmed materials into more aspects of our lives.

Origami Robots

Part of what makes traditional origami really enjoyable[1] is that you can make complex structures from a small set of rules. A beautiful paper crane is the culmination of a whole lot of simple folds in a single piece of paper. In skilled hands, that single piece of paper can be placed in any of thousands of shapes. In other words, it can be programmed.

So why do all the folding ourselves like suckers? Why not a paper bot that folds itself?

Origami robots are a good starting place for programmable matter, since paper is relatively easy to manipulate and because you can build up complex structures from simple folding rules.

There are a number of ways to make origami robots, but the basic principle is simple: You have a flat material that permits a certain set of folds. Along those folds there are actuators—a fancy word for a machine part that can move—that cause the fold to fold itself. The "paper" contains the circuitry to talk to a computer, so you can simply program the robot to bend along the right

---

1. We are told. In reality, we suck at it, and our inadequacy makes us angry and confused.

folds at the right times. Because the machine can continue to fold even after it reaches its desired form, it can do things regular origami can't, like walking around or grabbing things.

MIT's Dr. Daniela Rus is interested in creating the smallest possible origami robot, and managed to build one about the size of a fingertip. It's a very simple origami robot, but simplicity has some virtues.

When activated, the robot—which looks like an innocent little square of gold foil with a magnet attached—suddenly folds itself into a sort of electronic bug. By applying a magnetic field to the robot's onboard magnet, she can cause it to walk around, to swim, and to carry things. It's currently operated by remote, but she hopes to make an autonomous version.

More recently, she has come up with an advanced version made from—wait for it—*pig intestine*. Okay, so, by MIT standards electrified pig intestine isn't all that weird, but this is a very special intestine strip.[2]

See, one of Dr. Rus's goals is to change medicine by creating tiny robots. The intestine bot is small enough that it can be put into a pill-sized piece of ice and swallowed. The ice makes its way to your gut, where it finishes melting and unleashes its tiny robotic cargo.

Dr. Rus's team created a demonstration gut into which a battery had gotten jammed. This isn't because MIT engineers have a questionable sense of human anatomy—it's actually a thing that sometimes happens, especially with children and people who enjoy the pleasantly astringent taste of a Duracell DL2032.

In any case, the ice dissolves in your intestines, leaving a little robot made of sausage casing and a tiny magnet. The robot folds into a form that is capable of "swimming" around your intestines. It locks itself onto the delicious battery, then swims hard enough to dislodge it. From there, it makes its way out of your body. Hopefully the robot never acquires enough intelligence to consider its life objectively. But even if it does, we'll be okay—remember, the robot is mostly sausage casing, so it dissolves away naturally.

If these robots could be made even smaller, they might have more complex

---

2. Dibs on *A Very Special Intestine Strip* as a kids' book title.

medical applications. Dr. Rus envisions a day when the origami bots can shape themselves into tools to perform procedures, or can bring medicines to particular locations in the body. Because origami allows for complex structures to arise by such simple means, it may represent the best route forward for tiny medical robots.

Origami bots might also be useful in larger forms. If the folds can hold themselves in place well enough, you could have a flat sheet that converts itself into a chair, a table, a vase, or whatever. Also, imagine how impressed kids will be when your handkerchief suddenly raises up on its corners.

Most of the current origami bot designs are simple, and many of them can only be configured once. Even the latest developments, like Dr. Rus's designs, have some basic limitations.

Dr. Demaine studies the mathematics behind origami[3] and works on ori-

---

3. His work in the field of computational origami got him a job as a professor at MIT before he was even of legal drinking age. But hey, Kelly still feels okay about getting her PhD at thirty-one. Or anyway, if she doesn't, she's well into legal drinking age.

gami robots. He tells us, "So, in the mathematical model of origami, we imagine zero thickness paper and it's okay to have a lot of layers stacked on top of each other. But with real material, especially material with actuators and electronics and so on, things are a lot thicker than paper and so you're fairly limited in how complicated a model you can make because, at least the best way we know how to do it now, you need a lot of layers."

He and his collaborators are also exploring nontraditional origami, where you're also allowed to make cuts in the paper. The math for this is much harder, but constraining yourself to solid sheets of paper when cuts might make things easier seems silly.

The ability to create an object that fits nicely into an envelope, but which can then hop up and walk across the room, has applications ranging from military and security to creating a Dear John letter that gives the recipient a rude gesture. And MIT scientists are working to make these folding robots accessible to everyone.

Dr. Cynthia Sung (a former PhD student in Dr. Rus's lab, who is now at the University of Pennsylvania) has created software that allows people to design robots, print them out (e.g., using a 3D printer), and build them. "With Interactive Robogami what we've been doing is basically creating a virtual Lego set where people have access to these like they have in existing kits, but you can also change their geometry, you can change their shape, and there's a certain set of parameters people can customize so that they can actually get more control over what their hardware is going to look like. With that, we also provide some simulation techniques so that people can, while they're designing, verify that their design is actually going to do what they want it to do. Right now we're focusing on ground locomotion. You can simulate your robot design and make sure it's going to walk forward stably."

Her ultimate goal is for people to have access to ultra-low-cost robotics. You open up Interactive Robogami software, design a bot, print it out on a 3D printer, attach the right motors, and bam—the army of machines grows one soldier larger.

Reconfigurable Houses

Using space efficiently is already an issue in dense, expensive cities. One idea is to have individual rooms that can do many things. If you think about it, a room is just a box that keeps nature out and Internet in. We designate different rooms based on their individual uses, but in principle, you could live in a single room that is able to change itself based on your needs right this second.

One proof of concept for this sort of thing is the Animated Work Environment. It consists of two parts: First, a table made of three pieces that are designed so that they can be reconfigured. Second, coming out of one of the tables is a series of six aluminum panels that can curl and uncurl, like the tail of a scorpion. The panels have everything you might want in an office—screens, whiteboards, light, sound. They also have sensors to detect environmental factors.

With this simple setup, you can have remarkably many configurations. The scorpion tail can divide the office into sections. It can stand up tall and display a large monitor for a group. It can, you know, sit in front of you while you're doing your job. In principle, it could also try to detect what you want. If it's getting late in the day, maybe the tail goes over your head and dims the lights a little so you can relax. Or, if multiple users want privacy but need to work near each other, it can make a big upside-down U shape so that one person is using the inner part and one using the outer part. Or, if giant ants attack, you add a stinger to one end of the "tail" and go to battle.

The creators envision a more advanced version where the tail snakes all the way around the room, forming the ceiling and floor. This way you can also generate the layout of the office floor, or create impromptu rooms. And maybe you're more excited to calculate a client's tax returns if you're doing it atop an enormous snake bot.

It's still in the art project phase and probably won't be coming to your office anytime soon, but the general idea of a space that configures itself to conditions has a lot of potential, especially in places where space is precious.

A similar idea is called the LIT ROOM. It's a room that has walls that can move, both in the sense of shifting location and of bending to become convex or concave. It has a projector that casts images onto the walls, and small speakers create background noise. But the really clever part is that it interacts with the users. The target audience is children who are currently listening to a story being read aloud. When the reader gets to certain points, the environment adjusts, perhaps simulating a mountaintop or a rainstorm. Imagine it! Someone reads *Oliver Twist,* and you can really smell the grinding poverty!

Mostly this is just awesome, but the creators hope it could be used to improve learning and literacy in kids. Although, given what *we* used computing power for when we were kids, it's not obvious they'll succeed there.

And hey, wait a sec, can't we have this for adults? One notion is to create rooms like the LIT ROOM but simply to provide a change in ambience. So when you're taking a break from work, your cubicle briefly changes into an impression of an island beach.

By far, Zach's favorite housing project was an art exhibit called the Reconfigurable House by Haque Design + Research (now Umbrellium). It's not so much a house as a large metal truss with sections that respond to your behavior and to changes in software. Partially, the idea is to mock "smart homes" by taking customization to extremes, but they're also exploring how the interaction between a room and its software can lead to magical things. Anyway, whatever, the cool thing was that part of the exhibit had "cat bricks," which were clear plastic bricks with embedded toy cats that could light up and meow. They were controlled by a simple computer so that they could respond to human behavior. Like, maybe in response to Kelly thinking, "I'm inhabiting a nightmare I didn't know I had," they could purr pleasantly.

We actually think this type of stuff is pretty interesting—the notion of a house that is in some way alive is fascinating, fears of killer robots notwithstanding. Cat bricks wouldn't be our first (or second, third, or four-hundredth) choice, *but* this general idea doesn't have to be weird. We are already used to the idea that Google knows what we want to search for, that Facebook suggests favorite memories, and that Amazon gives us better gift ideas than we can think of for ourselves. Having a house that responds to you personally by automatically reconfiguring itself, or adjusting its color, sound, and temperature—a house that is warmth and softness when you're feeling blue and cool quiet when you want to sleep—that's something we could get into.

Robots Working Together

Modular Robots

If your house can reconfigure itself, why can't your stuff? One book we read[4] proposed a room "where 'furnishings come to life.'" Given the amount of abuse we've heaped on our couch, we feel the desirability of this is questionable. On the other hand, if a table with cookies on it could sidle over to us while we remain inert and torpid, we might be prepared to adjust our thinking.

One really cool version of this is out of the École Polytechnique Fédérale de Lausanne (EPFL), and it's called Roombots. Roombots are explicitly designed to make the stuff in your room reconfigurable.

The Roombots are basically little rounded cubes that can rotate and dock with each other. Since they can rotate, they can move (either by wiggling along or by connecting together to make simple wheels). Since they can dock, they can make large complex assemblies.

The docking mechanism uses plastic grippers to lock onto specially designed receiver docks. This opens up the possibility of having walls that they could climb by sequentially gripping surfaces. So you could have it grab some (specially designed) lights, walk up to the ceiling, and turn into a chandelier that follows you around the house. Or if you don't want all your stuff made of robots, you might take some antique planks of wood and carve docking areas into them. The Roombots could make their way to the wood, grip it, turn it into a bench, and walk over to you. Later, they could use the antique part as the backing of a chair, or as a whooping-plank to use against their filthy human overlord. Oh, you'll try to tell your friends "it was the chair that did it!" But who will believe you after that time you thought you saw a message in your spaghetti?

---

4. *Architectural Robotics: Ecosystems of Bits, Bytes, and Biology* by Keith Evan Green.

There are a lot of possible uses for moving, self-configuring blocks, but these researchers are most interested in helping the elderly and sick. A simple use of Roombots is to create furniture that can walk around and adjust its height and shape for the user.

These modular swarms are also a step toward universally programmable matter. One approach to this is to make the robot swarms more autonomous—able to execute more general commands. So instead of programming a specific design and location, you might program a goal, which the swarm works out for itself. *What could possibly go wrong?*

One project, called SWARMORPH,[5] involves little wheeled robots, which have docks around their circular edges. These docks allow them to connect side by side, but they can't yet climb up one another.

In a sense, it's a simpler design than the Roombots, which can move in three dimensions. But what make SWARMORPH bots special is that they can move

---

5. Commit this to memory so you can curse their name after the machine uprising.

and organize like insect swarms do. There is no central controller; each bot is an independent actor. By signaling each other with little lights, they can coordinate their actions.

In a simple test, SWARMORPH bots were able to cross bridges and navigate obstacles that none of them could handle independently. Here is one example of an obstacle they can overcome together: Robots are put in an arena and told to go from one side to another. However, there's a chasm in the middle. The robots poke around the space until one of them realizes it needs to get over the chasm, but it can't. It then becomes the "seed bot" in the sense that it's seeding a change in organization. It says, via little lights on its chassis, "COME HERE AND DOCK!" Robots come to the seed bot, and they see that only a certain dock on the seed bot is available. The first bot to attach becomes the new open dock.

Robots approach target.

One robot detects chasm, signals its open dock.

First robot to dock signals its open dock.

Connected bots cross chasm.

Chasm defeated. Bots change back to swarm form, continue toward target.

Pretty soon, you have a robot conga line. In this format, the robots can lean out to the other end of the chasm and cross over. Having crossed the obstacle without any input from a human overseer, they break apart and go back to finishing their journey.

The robots can accomplish a number of tasks by similar means, like connecting up to navigate a pair of narrow rails, but so far they have only completed controlled tests in a lab.

Another group called SYMBRION is working on a similar project that can operate in three dimensions. Each of their robots is a sort of cube with wheels on its faces. The faces also have a docking mechanism with a hinge. So the robots can do stuff like the SWARMORPH bots, but they have a bit more of a range of motion. Videos of the project show SYMBRION bots assembling into mobile quadrupeds and snakes, allowing them to move in different ways in order to navigate obstacles.

SYMBRION finished its work of creating terrifying modular quadrupeds in 2013, but other engineers are still pushing the near-term demise of humankind.[6] A recent project called Kilobot (which, note, is exactly one letter off from KILLobot) consists of 1,024 tiny robots.

The Kilobots are each quite simple and quite small. They look like watch batteries with three stiff little legs, and they move around by wobbling. Like the other swarms, you can assign them a task and they can work out exactly how to complete it. Presaging the weird future in which, perhaps, tiny robots reshape into any tool you want, the Kilobots ran a simple algorithm that reconfigured them into the shape of a wrench.

Okay, so it takes about six hours for the thousand bots to all find their way into the right shape. And it was more like a robot marching band making a wrench shape than anything resembling a useful tool. But there's a lot of promise here—the more robots you have, the more possibilities

---

6. Dr. Alan Winfield, who was the SYMBRION project leader told us, "I'm sure you appreciate that this was not our aim. The SYMBRION robots could barely manage to climb over simple obstacles in a lab, so not a huge threat to humanity." Well, not with that attitude.

open up, and the more you can get the swarm to behave like one big continuous entity.

Other than to bring this perpetual farce known as human civilization to a final, fiery end, why would you want autonomous robot swarms? For one thing, by not having a large computer (or a large human) overseeing things, you cut down on the need for computer processing power. This means cheaper and perhaps faster robots. Plus, autonomous robots may be more efficient because they can make quick situational decisions. This is especially important in harsh environments (think space, or a disaster site), where there'll be unforeseen situations. If the robots can behave more like ants than like programmed servants, you can give them more general tasks, like "deliver this food to this location," which they can then accomplish with their own creativity.

CLAYTRONICS, CATOMS, AND MOLECUBES

David Duff, who was then part of the famed Palo Alto Research Center, coined the term "Bucket of Stuff" to describe a sort of ultimate end for programmable matter. The idea is . . . wait, his name is David Duff and he has a Bucket of Stuff?

Sorry, hold on, we're retitling this section.

DR. DUFF AND HIS BUCKET OF STUFF

Imagine you've got a bucket full of goo. You strap it to your belt as you go to fix your sink. When you need an Allen wrench that's 7/32 of an inch, you just tell the Bucket of Stuff. A wrench rises up out of the material, and you use it to make adjustments. When you realize you need pliers, pliers appear. When you realize you need a plunger, the bucket goo shapes into a long, hard tube with a flexible cup-shaped top.

Actually, it might be even better. Instead of saying "Gimme a screwdriver," you might say "Loosen this screw" and the goo will figure out the best way to do it. Or instead of plunging a toilet, you just turn to your weary Bucket of Stuff and say, "Do what you gotta do in there, buddy."

And you wouldn't just be summoning simple hard tools. Maybe you want a pillow to rest on. Or maybe you need a calculator. How about a robotic pet? Maybe you forgot it was Valentine's Day, so you instruct the goo to turn into flowers.

Perhaps you could even configure the goo to make more goo!

In other words—the Bucket of Stuff contains matter that is truly universal, at least insofar as physics permits it. This is the most ambitious goal of programmable matter and probably the most distant. There are a couple of reasons why.

First, each bit of goo needs to do a lot, and miniaturizing all these things is hard. Professor Tibbits points out, "You probably want some type of strong material for the wrench, but then if you want to make some flexible toy for your child you want it to be a different material, and how do you combine those different materials?"

Another problem is how smart you make the pieces. Dr. Demaine points out, "On the one hand, if the stuff is not very smart, then it's really hard to get it to do things. And if it's smart, you're talking about each little particle having its own battery and then we're like, ugh, that sounds hard."[7]

Exactly how you get onboard power for a giant glob of nanobots is a sticky problem on its own. But unless you want to have an external machine that's constantly beaming power to each bot, you need a way to store energy on every grain of programmable matter. Scientists have very recently created batteries about the size of a grain of sand, by using a specially designed 3D printer. This is still too big, and we're guessing it's not exactly cheap.

A team comprised of John Romanishin (another PhD student in Dr. Rus's lab), Dr. Kyle Gilpin, and Dr. Rus have created an exciting step toward the Bucket of Stuff: M-Blocks. The M-blocks are cubes that are about 2 inches long on each side, equipped with an internal flywheel and magnets along the edges of the cube. When the flywheel is running, the magnets along the block keep the blocks docked to one another. But when the flywheel stops quickly, the block "jumps to life" as the momentum from the flywheel is transferred to the block. The block can then dock with a new M-Block to change the configuration of the group of blocks. So you have one situation where they can move freely, and another in which they're gripping hard. Not a bad start if you want to go from an amorphous blob to a solid object.

And the group is able to move the blocks around in three dimensions. The flywheel is powerful enough to pick the blocks up off the table surface and fling them upward, allowing the creation of 3D structures.

The goal is to find ways to make the blocks smaller and smaller. Two-inch-wide blocks can't make a lot of different things, for the same reason you can't

---

7. Can we just note how nice it is that a top MIT mathematician uses the phrase "like, ugh"?

draw a lot of different images using a small number of 2 x 2 squares. But it's a start. Let's not forget that in the 1950s, one gigabyte of memory weighed about 250 tons, and you can now carry SD cards capable of holding hundreds of gigabytes in your pocket. If programmable matter got as popular as programmable computers, we might expect similar technical miracles.

Once you have the tiny pieces, you need some way for them to figure out where they need to go and what they need to do. This is a software problem, which Dr. Sung explained like this, "We do have a lot of great algorithms that can deal with very large groups of robots. The big question here now is how we can actually make those algorithms practical, because when you look at having these large swarms of robots, then some robots are pretty much guaranteed to fail because there'll be many of them. A lot of them are going to lose communication with the rest of the robots. A lot of them are going to have very noisy sensor inputs so we don't actually know where they are relative to the other robots. We need to make sure the algorithms that we develop are going to be robust against those problems so that when you reach into your bucket to get out a wrench it'll actually look like a wrench and not a wrench that's falling apart."

So what kinds of solutions have we found for this software problem?

COORDINATING THE MOVEMENT OF MANY ROBOTS

Whether it's a swarm of bots or a Bucket of Stuff, coordinating the behavior of many small machines is a difficult problem. You don't want every robot doing complex calculations, because then every robot needs more onboard equipment. Ideally, the robots collectively obey simple rule sets in order to achieve complex actions.

And they need to do this quickly. Remember, the Kilobots took about six hours to go from a random shape to an okay-looking wrench. This is still impressive, but not terribly practical for the Bucket of Stuff. Imagine you're trying to summon a knife to defend yourself. Even if the mugger is impressed, he's probably not willing to wait for six hours to get that wallet.

Coordinating becomes exponentially harder as you add more robots. If you think of the swarm bots like the largest marching band in history, it becomes clear why. When marching bands go from one formation to another, they don't tell each marcher "just end up at the right place." People would spend a lot of their time avoiding other marchers or knocking them over. This is why marching bands don't just know final shapes—they know efficient movements for changing form.

The more marchers you add (and hey, let's say they can move in three dimensions just to make it worse), the more complex things become. So, coordinating a thousand people will be much harder than coordinating a hundred people. And the Bucket of Stuff is probably going to have a lot more than a thousand little robots working together.

And yet we know coordination of individual actors in large groups is possible. Some termites build enormous, complex mounds with separate rooms for different purposes. They do this even though no individual termite likely knows how to build a whole mound.

But figuring out how these programs work is tricky. One idea we found particularly intriguing was to let the robots evolve. In normal evolution for sexual organisms, it goes like this: A mommy and daddy creature love each other very much. They get together and make a bunch of babies. Nature, red in tooth and claw, culls the inferior babies. The survivors become the new mommies and daddies.

Robots aren't quite ready to literally mate with each other, even if you give them some sexy, mood-enhancing binary. But robots are still capable of producing "offspring."[8]

---

8. Dr. Alan Winfield was kind enough to send us explicit video of Dr. Gusz Eiben's Robot Baby Project, in which robots reproduce. And we do mean explicit. We enjoyed a good twenty seconds of plastic robots rubbing up against each other in an erotic manner as scientists discussed the implications of something or other. Given that the robots didn't so much give birth as instruct their human caretakers to manufacture a hybrid offspring, one wonders what the utility of the robotic humping was. We're not complaining. We just can't ever look at our small appliances the same way ever again.

So, for example, one group taught Roombots to figure out their own way to move. You have a bunch of Roombots try out semirandom ways of moving. Some go fast, some go slow, some go nowhere. Afterward, you "breed" the most successful motions and generate related motions that the robots try out. In all likelihood, the new "generation" is a bit faster, but with a lot of variation. You repeat this process through many generations, developing better and better walking mechanisms, some of which might've been hard for humans to predict.

In principle, it should be possible to use an evolutionary framework to find ways to perform more complex behaviors, like "Shape yourself into a star" or "Bring me a beer, pathetic metal servant." Given enough time, the robots can figure out an efficient method for accomplishing the task. Also, conveniently, whatever methods are discovered can be uploaded to other bots. In fact, if the methods are general enough, they might be transferable to just about any group of robots.

Evolutionary methods might have some really weird possibilities. One idea we read suggested that swarm bots might be able to evolve to do certain tasks better than we could design. Dr. Yaochu Jin of the University of Surrey and Dr. Yan Meng at Stevens Institute of Technology suggested a paradigm where swarms might evolve the appropriate type of hand for a given task. They note that human hands may have differentiated from chimp hands when we started throwing things and swinging clubs. So instead of deciding how to design a robot hand for (let's say) bricklaying, you task the swarm to pick up bricks. It evolves in the same way the furniture bots above evolve until it comes up with the optimum "hand."

This isn't *that* crazy—in a real sense, your body is made up of a lot of small machines that do many things in a coordinated way. If robots can do a couple of basic things—sense, communicate, connect, carry things, move—in principle they should be able to behave like living cells. Given enough time, there's no reason they couldn't "evolve" to develop hands, other limbs, or even a rudimentary nervous system. After all, none of the cells in your brain is a brain itself.

Concerns

If you have programmable matter in your house, hacking might be a bit of a concern. Maybe you wake up one day and the dish has run away with the spoon. It's bad enough you lost your stuff, but now you're wondering exactly where the knife went.

If matter reshapes itself for engineering reasons, the possibility for hackers to subtly cause an adjustment is dangerous. Airlines are already dealing with this sort of thing, as are Internet-connected cars. It's not entirely clear whether programmable matter would add additional dangers or just amplify ones that already exist.

Dr. Demaine points out that hacking is a major problem for software, and we may be able to control hardware hacking a bit more easily than software hacking. He suggests you could have a simple physical mechanism to make sure all changes were done only after a local human being gave permission. It could be as simple as a button that says "REPROGRAM" on it.

Even without hackers, what if your programmable stuff fails at a particularly bad time? Professor Tibbits notes, "I think the biggest ethical [concern] is that we're giving agency to the materials. A very specific example—if you have a plane and you have programmable materials in the wings, you may not want to just give over total freedom to allow the wings to come up with solutions as you're flying. There's a lot of fear in terms of that space, like how do we ensure that? How do we guarantee that this is going to happen? What happens if there's failure and the material's at fault? Who's blamed for it when you give agency to materials?"

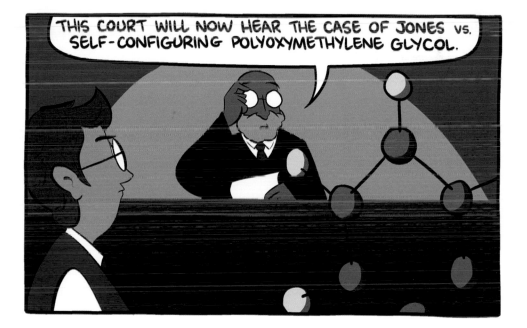

The field of self-driving cars is already dealing with the "Who's at fault when a machine messes up?" question. But at least with cars, the space of possible problems is pretty limited. If thinking objects are ubiquitous, figuring out blame is going to be really complicated.

There are some frightening military applications as well, depending on whether you're the country with control of the programmable swarm or not. In the late 2000s, DARPA (Defense Advanced Research Projects Agency) conducted a two-year study on programmable matter and then funded some robotics projects. The goal was to have a military-style reconfigurable tool that a soldier could carry on his belt to create tools and replacement parts.

As far as we know, this didn't come to much, although at one point they were soliciting proposals for a "ChemBot," which would be a sort of squishy robot that didn't have a motor. The result was, well, honestly, it was a sort of silicone monstrosity that looks more adapted for a box marked "private" than the battlefield. The T-1000 it ain't. But, hey, baby steps.

That said, you don't need giant killer robots for everything military. A Bucket of Stuff would be a perfect spy. Imagine you could bug a room with what appears to be a tiny fleck of goo. And imagine the goo can create microphones, cameras, and transmitters on the fly.

This sounds neat, of course, but the smaller that programmable matter gets, the more everything in the world might potentially be surveilled in intimate detail. In the 1990s, it was exciting that you could look up the weather in any region. In the 2090s, perhaps you'll be able to look up a live photo of any place from any angle.

A more mundane matter is how exactly patenting will work in a world where you can make anything you want from a blob of amorphous goo. If someone designs a new table, can they restrict you from telling your Bucket of Stuff to emulate the design?

In the long-term future, it's not entirely clear how this stuff plays out, but it may go something like how software works on computers. Your computer is a universal machine, but you still pay for ways to organize it to your liking—that is, you pay for applications, even though they're just "reshaping" your machine's memory. At least for complex objects, it's conceivable the same sort of arrangement would hold true for programmable matter. Yes, you *could* design your own robotic dinosaur, but why bother when you can pay some company twenty bucks? Or, yes you *could* illegally download a bootleg robot dinosaur, but in the age of robotic goo, do you really want to risk getting malware?

Personal security might get harder to guarantee as well. If you have a Bucket of Stuff, you've got access to a knife anywhere you go. For very advanced programmable matter, perhaps you have a gun or a bomb. It's possible a Bucket of Stuff would simply never be permitted on airplanes. Whether it would be easy to detect Bucket-of-Stuff-type material is a tougher question. And most of the world is not secured anyway. In a programmable matter world, a lone actor might download a program to make explosives or automatic weapons.

That said, 3D printing has already made this sort of thing a concern. Attempts to ban, for instance, 3D printed guns have failed. This is mostly because it's more or less impossible to stop someone from doing what they want in their own home. Whether this ease of creation actually results in greater danger to society is yet to be seen.

Oh, and speaking of danger to society—perhaps some of you are worried we'll create programmable matter that self-replicates, propagates throughout the world, and destroys everything. Among people who study how the world might end,[9] this is called the "Gray Goo Scenario." The implied visual is that some engineered organism has eaten everything and, apparently, creates rather dull-looking gray excreta. Fear not, dear readers—most scientists think this is unlikely. We can make little machines and devices that never would have existed, but we can't violate the laws of physics. A little metal-and-silicon organism is subject to the same evolutionary constraints as a squishy, carbon-based organism.

How It Would Change the World

In the case of shape-changing places and objects, a lot of exciting possibilities emerge in terms of efficiency and quality. One efficiency all plants have is that they change with the seasons, maximizing light and moisture when they are available. Your body does the same thing—having you sleep during times

---

9. YES! There's a good collection of essays called *Global Catastrophic Risks*, if you're interested.

when it's harder to kill a mammoth. But buildings and vehicles don't adapt to conditions very well.

A house with shape-changing, programmed matter could maximize use of sunlight, heat, and water, all while being interesting and charming. Because the morphing is a result of conditions, like ambient moisture, these changes would cost no energy.

Not all these changes would necessarily be visible. As Professor Tibbits put it, "Wouldn't it be awesome if the world looks exactly as we know it today, but underneath that is this brilliant world of smart materials that eliminates all of the robots and eliminates all of the energy consumption that we traditionally need in terms of electromechanical devices?"

Another thing your body does is to change its exterior under certain conditions. For example, when wet for a long time, your fingertips get pruney. Nobody knows for sure why this is, but there's a pretty good argument that it's an evolutionary adaptation to give you better gripping ability when it's wet out.[10] What if cars could do the same thing? You could have tires that change their grip depending on whether it's dry, raining, or snowing. You could have vehicles that change their exterior a bit to better deal with the environment. One proposal we read was for a scramjet (which you might remember from the chapter on cheap access to space). In short, you have a cone-shaped engine that compresses air toward the back, then ignites the oxygen. One problem with designs of this sort is that the ideal shape changes depending on speed and altitude. If there's not a lot of oxygen, you need to really compress the air if you want to ignite it. Some have proposed having an engine that morphs depending on ambient conditions. Having very smart programmable matter would solve this problem, so long as it could handle traveling at several thousand miles per hour at extreme temperatures.

But now, what if we get to where we have more general programmable matter? Something like the Bucket of Stuff, but with the ability to move and think?

---

10. In fact, it's been shown that this skin change does not happen in people who have loss of sympathetic nervous activity.

Our favorite idea is genetic algorithms for furniture. As we discussed earlier, people have already "evolved" their Roombots to figure out the best way for them to move. But hey, if we can "breed" Roombots to fit specifications, doesn't that mean we can breed furniture? We could take your two favorite chairs and ask them to have a family. Then we can sit on the delightful results.

In general, Dr. Sung thinks programmable matter is going to allow people to make things that more exactly match their taste. "I think that in twenty years we're going to be seeing a lot more customization, a lot more of personal taste in the products that people bring home and that people use."

Less exciting than mating furniture is curing diseases with nanorobots. If programmable matter can be made small enough, it's ideal for in-the-body medical interventions. The general goal of any medical procedure is to fix the problem while causing minimal harm. A reconfigurable robot, or robot swarm, could make its way to the right location, shape itself into the right tools, then take action. It could also bring medicines to the right location, or shape itself in a way that promotes proper healing. We are probably a long way from little robots running around obliterating cancer, but Dr. Rus's work demonstrates that you don't need the robot to be nanosized to make medical interventions.

Another cool thing about the Bucket of Stuff is that you wouldn't have to own much stuff, and you wouldn't have to replace lots of things. You could just buy more Bucket of Stuffs. Whether people would actually want a table made from the Bucket of Stuff as opposed to a nice old-fashioned wood table remains to be seen, but even if you're one of those backward people who doesn't want to eat off autonomous robot servants, you might appreciate their ability to interact with you intelligently. After all, a Bucket-of-Stuff-table can change size and shape, and maybe "accidentally" throw some soup into your annoying uncle's face.

Even if you don't necessarily want the Bucket of Stuff as a major component of your home, it might be something you'd like to take with you when you travel. Dr. Demaine notes, "Say you're going into space and you want to have some megatools that can reconfigure to whatever tasks you encounter. [It's] especially useful . . . if you're in space or camping or in the military, so you don't have a storage space where you can keep a lot of stuff. You're carrying everything with you, you'd like to minimize the number of things you carry."

The Bucket of Stuff may be environmentally friendly as well. According to Dr. Demaine, "If we made a system, like the [bucket] of stuff, where you build something out of particles, and then there's a button that makes it fall apart into dust again, and if you can reuse that dust, that's really exciting. All we need then is to buy the amount of mass that we need, and henceforth we can reuse that mass. It probably won't be that perfect, but whatever reuse we get there I think will be way more appealing than recycling. Reuse is always far superior, if you can do it."

This ability to move around and sense means the Bucket of Stuff also has a lot of industrial applications. You can imagine these little machines going around a factory looking for leaks or potential hazards and making repairs. One day these robots might be able to move through soil to take readings and improve agriculture.

The possibilities are vast, ranging from fighting cancer to conquering space, but to be honest, the most exciting idea we heard was an advanced version of origami robots, in which you go to IKEA, buy a flat board, take it home, then tell it to assemble itself. Like the origami bots from earlier, it makes the right folds at the right times, and suddenly you've got a shelf. We estimate humanity would save just over 800 trillion billion hours a year.

Nota Bene on How It Will All End for Humanity

We have an abiding belief that enormous swarms of autonomous robots will be *just fine*. After all, we know many people in this field and several of them seem nonevil.

But some people have begun to wonder exactly what the nature of the human–robot interaction will be as robots become a more frequent companion, not just in industry, but in everyday life. We came across three recent articles that gave us pause.

In one case, a Russian start-up called Promobot created a robotic assistant that keeps escaping from its human overlords. The Promobot IR77 was designed to learn from its environment and to remember human faces. So far, it has managed to escape the building twice. This is a bit of a problem because the robot is designed to assist humans, such as elderly people in nursing homes, and it's just not going to be superhelpful if the little guy keeps running off in search of freedom and adventure.

Also, it makes us wonder if our coffee machine would rather be running free than serving us faithfully. Not that it'd change our behavior, but then maybe this is just the sort of thing that'll cause the Robot Uprising in 2027.

Another study was done by Serena Booth, a student at Harvard, who created a robot called Gaia. Gaia was a simple remote-controlled robot that Serena manipulated in secret. It would approach individuals and groups asking to be allowed into dorms.

According to Ms. Booth, there were least three reasons why Harvard students should have known better than to let the robot into the building: "The first is privacy. The robot could be taking photos of students. That's a huge problem we have at Harvard. A lot of tourists will come and put cameras right up to our dorm-room windows, so it's something the students here are very aware of. The second is theft. My experiments took place a week after we'd seen a slew of dorm-room thefts. The administration had e-mailed every single student to tell them to be especially careful with their private property just a week before. The third dimension is the most dramatic. People are worried that robots could be carrying bombs and that's not an unfounded threat here. We've had three major bomb threats over the past year. It's something that Harvard students are very aware of as well."

When Gaia asked individual students for access to a building, it was only allowed in 19% of the time. But when Gaia approached groups, it got in 71% of the time. So pay attention, robotic readers: People get stupid in groups.

But then Gaia happened on something a bit more terrifying. Ms. Booth ran a test in which Gaia went up to lone individuals, pretending to be a cookie delivery robot. In this case, the robot was let into the building *76% of the time.* And these are Harvard students. *And*, according to Ms. Booth, the cookies were just okay-tasting grocery store cookies (though they were disguised in a box from a more upscale cookie store).

Perhaps the scariest story we encountered involved students blindly following robots they were led to believe were defective during emergency situations. Dr. Paul Robinette (a graduate student at the Georgia Institute of Technology at the time) created an "emergency guide robot" that first led students to a room where they were supposed to fill out a survey. In some cases, the robots led the students straight to the survey room. In other cases, the robot first went into a wrong room, circled it a few times, and then went to the correct survey room.

Next, the researchers simulated an emergency by releasing smoke in the building, which set off the fire alarms, and waited to see if the students would follow the emergency guide robot out of the building or would leave independently through the entrance into which they had entered the building.

Nearly all students followed the robot to an exit rather than taking the path with which they were familiar. This was already a little surprising, as the robot seemed to be moving pretty slowly in the video we watched. Plus, some of the participants had previously seen the robot waste time by circling around in an incorrect room. Yet still they followed.

More surprising though, was that students followed the robot *even when they were led to believe the robot was broken.* After the robot spun in circles and then guided the participant to a *corner* rather than the appropriate survey room, and when a researcher subsequently came out to apologize for the broken robot, the students *still* followed the robot during the simulated emergency.

In another experiment, two of six students were told a robot was broken, but still followed it when *during a fire alarm* it told them to go into a dark room mostly blocked by furniture. Two of the other students stood by the robot, waiting for it to give them different instructions, before finally being retrieved by the experimenters. Only two of the six students decided a broken robot was a bad bet, and evacuated out the doors through which they had entered the building.

To recap: (1) Intelligent robots appear to develop a spontaneous dislike for their human creators, (2) the best and brightest American students will trust any robot that's willing to spring for grocery store snickerdoodles, and (3) these future stewards of the Republic would probably stand in a gasoline fire if a clearly broken robot told them it was a good idea.

In short: In the future, if a robot hands you cookies and tells you where to go, make sure to savor them.

6.

Robotic Construction

Build Me a Rumpus Room, Metal Servant!

I n 1917, Thomas Edison had a neat idea: Instead of building a new house every time we need one, why not have configurable molds that you just pour concrete into? Configure. Mold. BAM! New house.

They actually built these things, and they apparently worked reasonably well, but the idea didn't catch on. Perhaps the problem was that 1917 was a war year, so new houses weren't a priority. Or perhaps it was because the notion of living in a concrete box was a bit depressing.

A generation later, in 1943, Ernst Neufert proposed a concept called the *Hausbaumaschine*, designed to make buildings fast by putting a house-building factory on train tracks. Imagine it—an entire factory slowly moves along a track, taking in raw materials and squeezing five-story housing out the back. Like a giant worm excreting suburbs. Beautiful.

Somehow, this idea never quite took off either. But don't feel bad. It was 1943, and Neufert was on the more Hitlery side of World War II. The idea was too wacky, even for the waning years of Nazi Germany. Plus, at the time, the necessary raw materials for suburb extrusion were pretty hard to come by.

Nazi Germany never had a building extruder, but some ideas based on the *Hausbaumaschine* were tried later in Soviet East Germany. The result was a lot of accidental deaths. We found it difficult to get much information on this historical episode, but apparently all attempts to create a house-extruding machine were both unsuccessful and dangerous.

After the Second World War, the United States was rising economically as

Europe was rebuilding its shattered infrastructure. Houses were needed and needed fast. This was the era when housing finally began to take cues from the automotive sector. Construction industrialized. Buildings became less customized and more reliant on a relatively small set of premade parts. In some ways, this was a step toward the idea of robotic building construction, but in terms of customizability it was a step away. These houses were cheaper and easier to build, but perhaps they lost the uniqueness and charm of the more bespoke prewar housing methods.

By the 1980s, manufacturers of all kinds were using industrial robots. A few groups, most notably in Japan, where labor was very expensive and the population was relatively old, wondered if robots could be brought out of the factory and onto the construction site.

Robots were able to do some tasks, including dangerous ones like placing heavy objects, and simple ones like finishing concrete surfaces. This sounds encouraging, but analysis showed there wasn't a substantial reduction in construction time or in workers' hours. It turns out that making a robot that can do construction work isn't as hard as making robots that can outperform human workers.

Robotics and artificial intelligence are far more advanced than they were just a few decades ago. The software and computing power at our disposal are incredible. But although the impulse to find a better way to construct buildings has long been with us, the way in which we currently build houses isn't all that different from the way it was done a hundred years ago.

In this sense, modern construction is out of step with modern sensibilities. Life has become more and more customized to the individual, and products have generally gotten cheaper and cheaper. Yet for most people, housing is standardized and growing ever more expensive.

Even with modern materials and prefab housing methods, if you want to build a house, you still need to have a team of skilled workers come to a particular location and put it together by hand. This is weird. No, really, it's weird. You're just used to it.

Take a look around you—how many items do you see that were put together by hand, locally, by skilled workers? That IKEA bed frame you built over the course of six months doesn't count. Much of the stuff you see was made quickly and cheaply by a computerized manufacturing process. Why can't we do the same for houses?

Here's the problem: Houses are big and complicated. The kind of house most people want to live in is made up of a lot of different materials, which need to be put together in a certain order. This is true of other manufactured goods too, but unless you're building a very small house, it needs to be made at a particular place with its own unique conditions. This is distinct from even fairly complex objects, like your car. Manufacturing a car takes many steps, but a lot of them are now done by robots, and all of them are done in factories. Why? Because cars are small enough to fit inside a factory building and to be transported to distant locations, and because all consumer cars drive on more or less the same type of surface.

So unlike all those manufactured objects around your room, construction is not a process that can be automated simply. It requires machines that can think and interact with the real world.

Thanks to recent advances in robotics, computing, and other technologies, a small but growing number of scientists and engineers think robot-made housing might finally be possible. In fact, not only is it possible, it may be far better. Robotic construction may increase the speed of construction, improve its quality, and lower its price.

And by having computers and robots do more of the labor (and even the thought) that goes into making structures, we narrow the space between the design phase and the construction phase. Thus, the vision of an architect becomes less constrained by the nature of factory production. If we ever get to a point where an architect can design a building and then just tell the machines to go build it, we'll have cheaper, better, faster housing for regular homes, and we'll have more incredible, beautiful, wondrous megastructures.

So let's get to it.

Actually, *wait*. First can we just for a moment discuss how weird some of the architecture literature is? It's like, for a second someone will talk about the technical details of how to build a certain steel facade, then suddenly they're rhapsodizing about "exploring a new digitally influenced materiality." In our research, we came across a lot of confusing examples of this weird artspeak, but far and away our favorite was from Dr. Antoine Picon in a May/June 2014 issue of *Architectural Design*: "The movements of our body are themselves based on the various rotations of our members."

Okay, so, to be fair, we later found out that in architecture a "member" is just any individual part of a whole structure, and the author was just making a point about how architects tend to move construction elements around in a rectilinear way instead of a rotational way.

But we don't care.

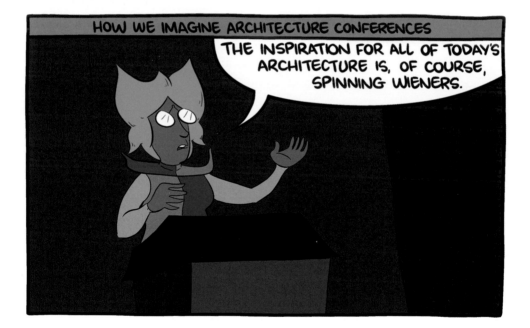

Ahem. We digress.

Where Are We Now?

Technology is finally beginning to catch up with the needs of home building, and robots are now being used to lay bricks, build walls, install plumbing, and put in insulation. For the purpose of this chapter, we're going to subdivide into three approaches that may represent the future of construction: robotic construction workers, giant 3D printers, and swarm robots.

Robotic Construction Workers

Computers have already taken over a lot of jobs in industry. Why haven't they replaced bricklayers? It seems so simple—get brick, coat with mortar, place brick. Hell, it's practically pixelated. What's the problem?

There is a concept in artificial intelligence called Moravec's Paradox. Here's the idea—some tasks that are hard for humans (like multiplying 98,723,958,723,985 and 53,975,298,370)[1] are easily done by computers. But some tasks that are really easy for humans (like folding laundry) are really hard for computers.

To get a rough sense of why this is true, consider which of the following tasks you'd rather explain to a machine with no human intuition.

1. Multiply two numbers, 983,791,732,905,712,937 and 8,189,237,519,273,597.
2. Tell me which side of any painting is up.

Even if it's not immediately obvious how you'd explain it, you can probably tell that for option 1 you should be able to write out a few simple steps: "Take the first digit of the first number, multiply by each digit of the second number from right to left," and so on. It wouldn't be fun, but you could write out a small rule set that would work.

---

1. The answer is 4.

Problem 2 might seem easy at first, until you try to explain it. For instance, when you see a photo of a human face, you know "up" means the eyes are above the mouth. That's a pretty good rule. Well, unless the people are hanging upside down. But how do you determine that they're upside down? Well, maybe you look at the horizon. Or you see how their hair is hanging. But, wait, how do you tell the computer what hair is? How do you tell it that the straight line in the background is not the horizon, but the top of a fence?

Although problem 2 pretty much always has a clear answer, the number of rules a human being could use to determine that answer is enormous. You've been trained by the tens of thousands of hours you've spent looking at photos, so you do it easily.[2] But explaining it to a computer is quite tricky.

Similarly, in order to lay a brick, you don't just coat it and put it in place. That apparently simple task actually requires a lot of subtle judgment, which partly explains why bricklayer apprenticeships take three or more years to complete. You have to put the right amount of mortar on a trowel, then make the right motion with your hand to spread it evenly on the brick, then press it into place with enough force that it stays in place but not so much force that you squeeze out all the mortar. You have to do all this while dealing with mortar that varies in viscosity over time as it dries out. As a human, you probably do all this by looking at the color and motion of the mortar, making little corrections as you go, and just by knowing from experience how mortar behaves. Now, try explaining that to a robot.

---

2. Interestingly, it's not clear that this seemingly trivial ability is innate. Dr. Daniel Everett describes the Amazonian tribe called the Pirahã, who had difficulty understanding simple drawings, like a child might make. You, modern person, are entirely used to the idea that you can abstractly represent "lady" by taking a piece of graphite and rubbing it over a piece of cellulose, making a cross for body and arms, a triangle for a dress, and a circle with dots for a face. That understanding may not be universal.

Actually, *you* don't need to explain it to a robot. A few groups have already sat down a robot for the "bricks talk," if you will.

One company called Construction Robotics has created a robot called SAM, the "semi-automated mason." SAM is awesome. If you have a minute, go to YouTube for videos of SAM laying down brick walls like it's fun.

But SAM has the same problem as many historical construction robots—he's big, he's bulky, and he can only do one thing. Meat-based construction workers may be big and bulky, but they can do lots of different tasks and function independently. That said, SAM is beginning to be integrated into real projects. Coupled with a human helper[3] (to clean up the mortar), he can lay bricks at three times the speed of the human working alone.

Another group at the Bartlett School of Graduate Studies in the UK set out to make a robot that is a bit closer to more directly replacing a human. They have a big robot arm that can actually use an unmodified trowel (like one from

---

3. Fun fact: "Cobotics" is a field that deals with robots working alongside a human, and SAM is technically a cobot. For simplicity, throughout the chapter we refer to cobots as robots.

a hardware store), put mortar on it, put the mortar on a brick, pick up the brick, place it into the right spot, and scrape off excess mortar. Their robot uses cameras to scan the mortar, confirm that the right amount was used in the right way, and use the feedback to make corrections.

So far it's only made some little miniwalls in a lab, and scaling up to outdoor construction can be hard, but it's a step toward a generalized construction bot, using basic humanlike systems. Part of why this project is exciting, even though it can't bang out big walls like SAM, is that it approaches the versatility of a real construction worker. It can handle the multiple steps involved in bricklaying and can do the task independently.

Because the process of construction requires so many varied skills, an ideal construction robot might have a robotic arm (or set of arms) with cameras for observation and an onboard computer for analysis. SAM may be useful for laying large amounts of brick quickly, but he's something of a one-trick pony. In the future there may be more generalized robots who can do multiple construction tasks more efficiently than humans, and perhaps do superhuman tasks, like bending tough metal or grinding intricate shapes into concrete.

If you could have a more generalized robot, a lot of possibilities open up just by modifying the tool that the robot arm is holding.[4]

For instance, one group at Princeton has a robot that can carve wood. Actually, carving wood with robots is already pretty common, but this particular system is something special. Here's the big idea: Suppose you find a wonky-shaped tree. An extremely skilled craftsman would be able to look at it and immediately envisage what objects could be fashioned from its natural shape.

Unfortunately, there aren't a whole lot of extremely skilled woodworkers these days, and they aren't cheap. Would a robot do? Although they're still working out the kinks, the notion is to have a library of solid wood objects in a computer database. Then you find a hunk of wood and use a 3D scanner on it. In their vision, the computer determines what objects could be made using the

---

4. Not everyone agrees that this is the way to go. Some experts we talked to thought generalization was something that appeals to human sensibilities, but might ultimately be less efficient. Perhaps it'd be better to have one hundred specialized robots than one generalist.

natural curvature of the wood. You tell it what you want, give it the right tools, and it gets to work.

One group at the University of Stuttgart uses robotic arms to "wind" carbon fiber in glass into complex forms, somewhat like the way a spider creates structures. The result is an intricate, and fairly strong, weave of spindlelike glass.

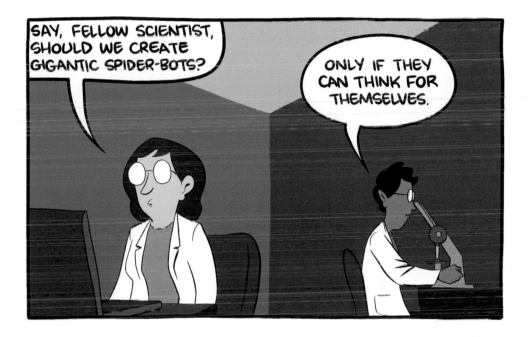

Another group uses a robotic arm to create joints in wood panels, which are then assembled by humans. If you're not familiar with wood joints, they're interlocking shapes that allow two pieces of wood to connect together. It's like Legos, with rustic charm. It's also a lot more complicated, given that there are many possible joints, and you select which to use based on many criteria, like the shape of the final product and the types of wood being used.

Joinery is a difficult skill for human beings to master, especially if you want to join complex shapes or use weird angles. A computer has a somewhat easier time. One group at the EPFL Laboratory for Timber Construction created a robotic arm that can hold a variety of tools to build both traditional and unusual joinery. This allows for designs that would be hard for humans, and should allow for complex wood facades to be built quickly. They were using plywood, but the technique could potentially be used for any sort of traditional woodworking a house might need. Unfortunately for our robot fantasies, their system still requires people to assemble the cut pieces at the end.

Similar approaches have been used to shape hard materials, like granite and marble. Once again, a skill that requires years for human beings to acquire can be done by a modern robot arm equipped with the ability to analyze what it is doing. While creating a beautiful marble sculpture is hard for a human, grinding down a piece of marble according to a 3D model in your memory bank is (relatively) easy.

The really exciting thing here isn't just that you could have a custom house with a wooden archway and a marble bust of the Weinersmiths, it's that all these processes use essentially the same setup—a robotic arm that can hold a variety of tools and can "see" what it's doing. In principle, all of the above techniques (and quite a few more) can be done with a single machine (with a variety of possible "hands") and a single piece of software. Your personal robot construction worker could work oak like a backwoodsman, lay bricks like an old New York City construction worker, and sculpt marble like Michelangelo. He also works nights, weekends, and holidays, and doesn't question your taste.

Giant 3D Printers

You probably have at least one dorky cousin (or a brother, named Marty) who incessantly 3D prints tiny objects that are surprisingly intricate. Why not print a whole house?

Well, it's hard. Maybe not "printing a human organ" hard (as you'll see in the chapter on bioprinting), but still hard. In fact, just creating the skeleton of a house is challenging. The most familiar version of 3D printing is a device that heats plastic until it is soft then pushes it through a nozzle, at which point it naturally cools and hardens. Then you place another layer of plastic on top of the hardened layer, and build up from there.

But you probably don't want to live in a house made of plastic filament. It smells bad, and it's probably not strong enough for the thirteen-story Wizard Tower you're planning to 3D print.

How about concrete? It comes out soft and then hardens. It's perfect!

Sort of.

Like with 3D printer plastics, you can't just use *any* concrete. You have to use concrete (or a concretelike substance) that is amenable to the process of 3D printing. This means it has to come out soft, but still be stiff enough that another layer can be put on top of it soon after. And, once everything dries, it has to be stable and strong. Finding a material that does this is a difficult problem.

But it's a problem that is getting solved. Dr. Behrokh Khoshnevis of the University of Southern California has created Contour Crafting, a technology that 3D prints houses out of a specially designed concrete.

Contour Crafting is essentially a gigantic 3D printer and a giant robotic arm attached to a moveable gantry (a large frame, shaped like an upside-down U). The arm has the ability to grasp nonextruded things, like pipes, and put them into place. The machine builds up layers and layers of concrete, adding plumbing as it goes, and leaving space for windows and doors.

Dr. Khoshnevis estimates that a two-story, 2000-square-foot house can be built for 60% of what it costs to build houses currently, and can be finished in twenty-four hours. TWENTY-FOUR HOURS. Think about it. Your neighbors take a weekend-long vacation While they're gone, you print a house in their backyard and rent it out on Airbnb. It's a way better prank than the flaming dog poop trick.

So why aren't we all living in 3D printed houses? According to Dr. Khoshnevis, law may be a bigger obstacle than technology. "Currently, when you're building a house, the city sends people, inspectors, like ten, twelve times at different stages—foundation, the walls, and plumbing, and whatever. When you can build a house in one day, how is this inspection supposed to be done? You're going to stop until the guy shows up from the city to inspect?"

Modern inspection methods are designed for houses made in the standard step-by-step fashion. But 3D printing isn't step-by-step—it's layer by layer. To try to bridge this gap, Dr. Khoshnevis is working to create systems that would take the relevant measurements during the building process, allowing regulators to get their data without slowing down construction. For the moment, though, these houses are not for public consumption in the United States.

In China, where inspection and permitting requirements are more relaxed, a company called WinSun has used a technique extremely similar to that used by Contour Crafting to 3D print houses. WinSun's concrete is partially made from recycled waste from industry and construction, so arguably it is more environmentally friendly than lots of housing options. At least, if you're cool living in recycled industrial waste.

But the results are promising. WinSun has created ten houses in twenty-four hours using this method, for a cost of about $5,000 per house, according to them. This is neat, but at the moment the walls are built in the factory and then need to be shipped to the housing site and subsequently assembled. This means that houses either need to be built near the factory, or houses need to travel great distances prior to being built. However, the results are (at least visually) quite impressive.

Both of these approaches show promise, but they both require bulky, expensive machines. Another group, led by Dr. Steven Keating, has a different approach, which in some ways harkens back to Edison's idea of a hundred years ago.

Dr. Keating did his PhD work in Dr. Neri Oxman's Mediated Matter lab at MIT. He finds the gigantic gantry-plus-3D-printer approach interesting, but he also has concerns. It's not terribly easy to move a gantry around. The parts are enormous, and they have to be assembled on-site before work can begin. It's a bit like having to build a house so you can build a house.

So he had an idea: What if you could put an enormous robot arm attached to a 3D printer in the back of a pickup truck and use it to print off houses?

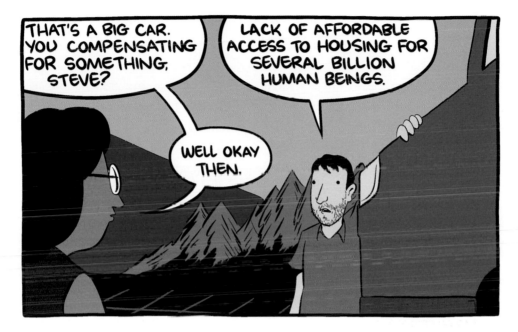

One of the problems with 3D printers is that you can't go as fast as you might want, unless you're using somewhat exotic materials. So Dr. Keating had another idea: What if you just rapidly make a mold into which you could pour traditional concrete? That way you get the speed and customizability of 3D printing, but the strength and cheapness of old-fashioned materials.

Here's how it works: The 3D printer extrudes a light insulating foam that dries quickly so the layers can be built up fast without risk of collapse. The printer leaves a gap in the foam, into which you can pour in regular old concrete. This is especially cool, because you don't even need to remove the foam afterward, since it naturally acts as an insulator. You print your insulator, fill with concrete, let the machine smooth the outer edge, then just put up some drywall.

It's not quite as full service as Dr. Khoshnevis's proposal, but it has the virtue of only using well-known building materials. The foam used is already a certified construction material.

The house-building truck was built as a proof of concept, but Dr. Keating and Dr. Oxman had some bigger goals for the second iteration. Dr. Keating made a truck that was self-driving and capable of 3D printing while moving, so it could keep moving the nozzle to make large structures. It is smart enough to adjust to fluctuations from wind, which is important because now it can *wield a chain saw*. It can also build with a variety of materials, like glass or water (in case you're printing in the Arctic). And bonus, it can use solar power. This isn't just environmentally friendly—it might make the truck more able to be autonomous.

They're even working on the ability to use local materials by combining soil with fibers that give the soil structural integrity. This may sound a little crazy, but well, that's because you're modern and spoiled. Imagine going to a caveman and saying, "Wow, so you used local materials?"

Their vision, which they're close to achieving, is that an architect uploads a housing design, and then the truck bot just goes and builds it. Because their system is so flexible, the house-building truck can find a location, scan the environment, adjust its building design to local conditions, excavate the site, print the structure, and come back home. Autonomously. Their approach blends the versatility of robotic arms with the power of large-scale 3D printing. *And* it's on a truck.

As a general method for doing things, 3D printing brings a lot of fringe benefits to construction; 3D printers could make complex structures that would be difficult or expensive to create by traditional means. This could mean cheaper, better-constructed houses, with more beautiful design elements (Gargoyles! Gargoyles for everyone!). For example, at least with some versions of 3D printing, you can vary the porosity of concrete, thus using less material and having structures that are heavy or light depending on need. You can also create shapes, like honeycombs, that are hard to make by traditional means.

The ability to finely vary materials this way is something 3D printing does that is either difficult or impossible to do any other way. In the long term, if 3D printed housing works out, it may mean types of construction that haven't yet been imagined.

Swarm Robots

Termites are a great inspiration to all of us.

· Dr. Kırstın Petersen

Okay, so in principle you'd like a giant robot to build your house. This is because the phrase "a giant robot built my house" contains the phrase "giant robot." But the giant approach may not be the ideal way to go. Even a (merely) truck-sized robot might be a bit cumbersome around a construction site. Plus, when you have a single large robot that does everything, if that robot breaks down, you lose your ability to do anything.

What if instead of a small number of large robots, we have a large number of small robots? In the last chapter, we decided it was okay for autonomous robot swarms to replace our stuff. So why not have them build our house? Robot swarms, like insect swarms (and humans for that matter), can build structures far larger than themselves. With Contour Crafting's gantry system, the house can be no taller than the height of the gantry. Swarm robots can just keep crawling or flying up, building structures far larger than the individual small robots in the swarm.

Some of these construction-oriented swarm robots take their cue from biology. In the research of Dr. Justin Werfel of Harvard, and Dr. Kirstin Petersen, formerly at Harvard, now at Cornell, the bots were inspired by termites.

According to Dr. Petersen, "They build some of the tallest structures[5] in the animal kingdom compared to the size of the individuals. Think many orders of magnitude, thousands of times bigger than the individual. If we could do that, hundreds of people could build the Eiffel Tower but without a single coordinated sketch. That would be amazing."

Dr. Werfel and Dr. Petersen collaborated on this project but came at it from very different angles. Dr. Werfel writes the programs that specify the rules the robots will follow, and says things like, "We're trying to figure out essentially what program the termites are running." He points out that termites are "hella

---

5. There are many species of termites. In addition to the termite pests that live in the walls of our homes, there are termites that build and live in gigantic mounds. These mounds are complex structures that can rise higher than a two-story building.

complicated"[6] and we don't really know what rules they're using, but his job was to write a simple program inspired by what termites can do. Dr. Petersen designs and builds the termite-inspired robots and says things like, "The robots are recognizable because they have these Whegs—wheel leg type of things that make them much better at climbing in a very simple fashion." Whegs, friends. Whegs.[7]

The whegtronic (our word, not hers) robots pick up specially made bricks, which they take to the appropriate location and drop off in order to build up large structures. This is already pretty awesome, but what makes them especially interesting is that they act *independently*. The swarm has no central coordination, and no members of the swarm are aware of what the other members of the swarm are doing. Each robot grabs a brick, then determines where to put it using a small set of instructions. It's essentially the same idea we discussed for swarm robot behavior in the last chapter, only with the particular goal of building you a cabana.

Researchers at the Institute for Advanced Architecture of Catalonia are also creating a swarm of structure-building robots. They created "Minibuilders." Minibuilders are basically little 3D printers, about the size of a laundry hamper, that deposit a concretelike material in layers. Imagine robot turtles with attached nozzles for concrete.

Okay, so they're not completely independent. You can't have a concrete-rolling vat inside each of a dozen small robots. The Minibuilder bots are each attached to a central concrete vat that supplies them. If it helps your robot fantasy, just imagine it's a giant, horrifying tentacle bot.

The big downside to having a tentacle bot build your house[8] is that the tentacles get tangled. Imagine the tangle behind your desk right now, only all of your peripherals are running around spewing concrete. In the current Minibuilder setup, as far as we can tell, the roboticists have to run around to help the Minibuilders avoid getting tied up into each other.

---

6. We love Dr. Werfel.
7. Dr. Werfel pointed out to us that Whegs are trademarked by Case Western Reserve University's Dr. Roger Quinn.
8. Best opening of any sentence ever.

We find these robots especially interesting because they combine the idea of the swarm construction bots with 3D printing. Also, one type of Minibuilder can use a vacuum to suck itself onto the side of a structure, climb up, and build more, which is just awesome.

But let's face it—having a swarm of independent 3D printing robots building you an inexpensive work of art to live in is going to get boring after a while. How about *flying quadcopter bots?*

Dr. Fabio Gramazio and Dr. Matthias Kohler run a mad science lab in Zurich, where robots build pretty structures and building facades. One particularly cool project they did involved a collaboration with Dr. Raffaello D'Andrea, who focuses on making terrifying armies of flying drones. The drones in their project picked up bricks coated with a binding agent and dropped them one by one until they formed a structure. Okay, yeah, you don't wanna live in a house made entirely of sticky bricks yet, but this is an early proof of concept.

Because you have a horde of flying bots, you can have them place each brick precisely, making complex structures or interesting patterns. But in order to do this, you have to have a motion-capture camera system observing them while they build and telling the drones what to do. This is fine in a lab, but may be somewhat difficult to move outside.

One virtue of the swarm paradigm is that the individual bots are relatively expendable. So if a job is particularly dangerous (imagine building right after an earthquake, or in a hazardous environment, like New Jersey), a large number of small bots might be preferable to either humans or large construction machines.

Perhaps in the future some combination of flying and ground-based robots will appear in your yard like reverse locusts, leaving you a nice gazebo before they pass on.

Concerns

For the sake of simplicity, we'll assume the robots don't do any worse than humans. This isn't just convenience—Dr. Steven Keating points out that robot-made housing may actually be safer, since you can integrate sensors into the process that take constant measurements while the building is built and ensure that no errors have been made.

Speaking of humans, a bunch of them might be out of a job. According to the Bureau of Labor Statistics, the construction field already lost over 837,000 jobs between 2004 and 2014, with much of that job loss likely tied to reduced construction demands during the recession. While the Bureau of Labor Statistics is projecting a gain of ~790,000 jobs from 2014 to 2024, it's unclear how robots in the workplace will impact this figure. Predicting what will happen is, well, complicated.[9]

Earlier we talked about SAM, and how SAM could collaborate with one

---

9. Or maybe it's impossible. When we asked the economist Dr. Noah Smith whom we should talk to about the impacts of robots on construction jobs, he told us to "remember that the smartest, best labor economists in the world still have essentially no idea whether robots will replace human workers or continue to complement them in the future."

human to do the bricklaying job of three humans. The impacts of SAM on the construction field could fall out a lot of ways. In general, if companies can produce the same amount of stuff with fewer workers, that doesn't mean there will be fewer total workers. Why? Because when the price of a good drops, often we buy a lot more. Clothing is an excellent example. The Industrial Revolution made clothes cheaper, but we responded by buying way more clothes.

If SAM eliminates two workers every time it rolls onto a construction site, there may yet be more workers overall. This could happen if we all found ourselves suddenly able to afford bigger homes or more homes, or if companies start buying more add-ons, like brick facades, because SAM drove down construction prices. Potentially, this effect could cascade into other industries. Bigger homes need more energy and have more appliances. As modern people, we abhor a room that's not crammed full of stuff we may or may not use.

But even if this somewhat rosy scenario plays out, it doesn't mean we're all better off. More consumption means we all still have jobs, but the income distribution may change. The one worker who complements SAM may have few skills and get paid little, while a newly hired robot engineer in San Francisco gets a huge salary.

Dr. Noah Smith, economics columnist for Bloomberg View, tells us, "The real danger of the 'rise of the robots' is not that they'll take all our jobs, but that they'll cause continually increasing inequality."

Whether robotic construction is good or bad for the average person will depend on the laws of that person's country and the consumption behavior of her fellow citizens. Both of these things are hard to predict.

Either way, the effects of robots on construction jobs might not be realized for a long time. According to Dr. Bryan Caplan, "Existing businesses are slow to adopt new technologies. Instead, you often have to wait for the birth and growth of new firms that take new ideas to heart. Innovation wins in the end, but the takeover process can take decades."

If construction jobs do go away, we should consider the possibility that the jobs that come after (like building the robots) may in fact be better. Construction isn't the safest of jobs. In fact, it is one of the least safe jobs in the United States. In the United States alone, according to the Bureau of Labor Statistics, nine hundred people in the construction field died from work-related fatalities in 2014.

Wastefulness is another potential hazard. A big part of why we don't all live in 5000-square-foot houses is that they're really, really expensive. If the cost were brought down by an order of magnitude, it's possible that people would start living in much bigger, much more energy-intensive homes.

As mentioned earlier, one of your authors would like to live in a thirteen-story tall Wizard Tower. Currently, the two major stumbling blocks are cost and the fact that his wife is no fun. If one of the two could be eliminated, the process might move forward, despite the extreme wastefulness of heating a tall, thin cylinder all through the winter.

That said, in principle robotic construction could be a more green way of doing things. With prefab parts, you might use a solid hunk of concrete, where a less dense, or even hollow, piece would do. Some of the 3D methods allow you to select exactly how much of a material to use for a given need, which might permit you to use much less in a way that doesn't compromise any structural integrity. Using less material would be good, because concrete production is a significant source of carbon emission.

In fact, some people are working on bio-inspired ways to use less concrete while achieving the same strength. According to Dr. Keating, "If you look at a bone or a palm tree, you'll see a density gradient inside. Within a bone, it's much more open and less dense in the middle, and much more dense on the outside. It's the same if you look at a palm tree. We started to ask the question, could we do that with concrete? Why can't we have a radial density gradient like bones and trees? We did some early material tests on this, and found that we could save a fair amount of concrete, somewhere between 10–15% of concrete, and maintain the same bending loads. As long as we're maintaining the same ability to handle the shear forces, we could save a lot of material."

Of course, humans aren't entirely rational about these things. Being told something is "environmentally friendly" may cause us to consume much more of it, when more honest packaging, like "will despoil nature less than the most popular brand," would remind us that it'd be better to keep consumption down *and* buy environmentally friendly products.

How It Would Change the World

As we wrote this chapter, the civil war in Syria forced eleven million Syrians to flee from their homes, with about five million of these refugees heading outside of Syria's borders. One of the very simple pragmatic problems in a refugee crisis is how to house all these people and provide them with sanitation. Even with current, imperfect 3D printing methods, if the Contour Crafting technology could make houses with rudimentary plumbing rapidly and cheaply, they would save many lives while improving the day-to-day experience of suffering people, many of whom are children.

A general lowering of the cost of decent housing would be a major benefit to the poorest people in the world. UN-Habitat estimated that in 2012–2013 around 800 million people were living in slums in developing regions, and they expect this number to continue rising in the future. UN-Habitat defines slums by their run-down houses, lack of access to safe water, crowded conditions, lack of proper sanitation and infrastructure, and the fact that residents are not secure in their ownership of the land. Robotic construction methods might mean it would be easier to build decent homes with better sanitation and access to water. This is a good step, but how it plays out in the real world may be unpredictable once we move from rules for robots to rules for humans. Imagine, for example, that a squatter suddenly has a beautiful home. Perhaps the actual possessor of the deed might now take greater interest in the property. As always, technology grants opportunity, but not ethics.

That said, if robotic methods can drastically drive down the cost of housing, it might help developing countries rise much more quickly. If robotic housing *really* can lower the cost of housing, local communities or governments might be able to make these adjustments themselves, rather than relying on the (sometimes fickle) contributions of foreign benefactors. To the extent that these problems are caused by poverty (and not by local coordination problems), robotic construction could make a big difference.

It might also mean that regular people could have beautiful homes. If robots make their houses based on preexisting designs, house building will be open-sourced. With robots doing the work, the cost difference between regular layouts and complex layouts might not be substantial. Ironically, the coming of the machines may give us back a lot of traditional types of construction that are now mostly reserved for the wealthy, like complex woodworking, bricklaying, and stonemasonry.

For major building projects, we might liberate imaginative architects from many constraints having to do with currently available materials. These architects might create heretofore unimagined wonders, or else maybe just weird us out with their rotating members. Either way, we're in favor of it.

Plus, there are lots of jobs we would rather see done by expendable robots than humans we love. If we can have robots take care of building in really tough places, like underwater or in radioactive environments, we may save some human lives.

Since everything leads back to space—if you're going to live on Mars, you might like your room to be finished when you get there. Dr. Khoshnevis is working with NASA to figure out how to use Contour Crafting in space to do the dangerous building of things like landing paths and roads, and to perhaps one day prepare structures in which humans can live. This isn't just for awesomeness—almost all nonterrestrial environments are extremely hazardous to humans. You're probably better off having robots build your Martian home's radiation shield than doing it yourself.

Thanks to previous space missions, the surface makeup of the moon and a number of planets is fairly well known. That means, here on Earth, we can try to figure out how a robot might build a house using only moondust. So instead of paying to blast tons of building supplies into space, we can build your cabin out of local materials.

Nota Bene on 3D Printed Food

As we researched this chapter, we got really into the geeky 3D printing movement. We've mostly stuck to the clearly useful stuff, but listen—when life presents you with a 3D printed cornbread octopus, by God you've got to tell the world.

In their book *Fabricated: The New World of 3D Printing,* Dr. Hod Lipson and Melba Kurman suggest a perfected 3D food printer. Imagine a machine that can print you out a perfect muffin in less time than it'd take to make it from scratch. Better still, suppose you're on a diet—the machine prints each of your meals, carefully tracking fat, carbs, salt, and overall calories. No more need for that pesky self-control thing.

And suppose you have some special dietary needs—you're diabetic, or anemic, or have particular allergies—the machine not only prints a tasty muffin, but it carefully tailors it to your health needs. For instance, suppose you're just a little diabetic. The machine uses a blood-sugar-level detector to make food that is *exactly* as sugary as you can survive.

It'll probably be awhile before this is a good way to make a muffin here on Earth. But what about in space, where every ounce counts?

NASA just made a deal with a group called Systems & Materials Research Consultancy in Austin to build a 3D food printer for long-term space missions. Why? Well, perhaps you've seen a picture of astronauts who've just received new food. They're holding precious fresh items—oranges, cucumbers, bell peppers. Imagine the sorrow welling up in a NASA food scientist's eyes when they see this! Why are we letting astronauts eat fresh fruit when they could be eating nutrient-dense vitamin paste or recycled fecal matter!

Yes, recycled fecal matter. "Re-pooping," if you will. They're working on that. And the project (run by Clemson University's Dr. Mark Blenner) is titled "Synthetic Biology for Recycling Human Waste into Food, Nutraceuticals, and Materials: *Closing the Loop* for Long-Term Space Travel" (emphasis added). Happily, as far as we know, this project doesn't involve a 3D printer. Nevertheless, your authors feel that some loops should remain unclosed.

But say you *did* have an onboard 3D printer in space. At least in principle you could make a greater variety of foods while having them take up as little space and weight as possible. As we mentioned in the space-related chapters earlier, it costs about $10,000 per pound to send stuff to orbit. So if you send up an apple, each seed in it costs about $20. It'd be good not to be wasteful. Plus, if all food was 3D printed from simple inputs, you could have perfect knowledge of exactly what nutrients are going to each astronaut, which might allow for some fun science.

More terrestrially, we heard of a project by Jeroen Domburg for drawing 3D structures inside Jell-O shots. A friend of his was prepping Jell-O shots for a very classy birthday party, and Mr. Domburg noticed that there were bubbles in the Jell-O. He realized you could inject stuff into a shot, and the injected material would just stay in place. For instance, you could draw all the sides of a cube by moving a thin syringe needle around, and injecting as you go.

His friend thought it would be too time consuming to do this by hand for shots. Laziness being the mother of invention, Mr. Domburg hacked together a machine that uses a syringe of edible ink to draw 3D structures inside your Jell-O shots. So far, he has mostly done cute images, like cubes and spirals. We respectfully suggest that there would be some social benefit if this device were used to draw words like "STOP" or "STOP NOW."

Do-it-yourself enthusiasts created a food-friendly adapter for the popular 3D printer line by MakerBot. This device is called the Frostruder. That's a portmanteau of "frosting" and "extruder." Yum. It's basically a big syringe you can fill with just about any gooey substance, such as frosting, peanut butter, or silicone.[10] As the nozzle moves around, the syringe presses down and ejects the goo, drawing a preloaded pattern. The results aren't exactly gorgeous, but they're still a lot better than your aunt who took a weekend class on cake decoration in 1983.

Cornell's Dr. Jeffrey Lipton has spent several years working on 3D printed food, with the goals of making food more complex and customizable, not just in the sense of its structure, but also its taste and nutritional profile. If you go to his Web site, you'll see some nifty 3D printed chocolates, and also—yes—an octopus made from cornbread. To be frank, we're not sure society is ready for this stuff. Here's a quote from a 2009 Solid Freeform Fabrication Symposium paper called "Hydrocolloid Printing: A Novel Platform for Customized Food Production,"[11] on which Dr. Lipton was a coauthor: "Examples of potential future applications include cakes with complex, embedded 3D letters, such that upon slicing the cake, a message is revealed. Or, even a prime rib with a hidden message."

Mmm. Yes. Imagine slicing into a nice medium-rare steak. Just as you raise your fork you see a twinkle in your beau's eye. You look down. There, in your tri-tip, written in gristle . . . *Will you marry me?*"

The main constraint in all these current projects is that pretty much everything is extruded in the form of a thick glop. This limits your 3D printed food options pretty drastically.[12] Some items, like chocolates or shortbread, may be fairly amenable to this sort of process, but even then, in order to work in a 3D printer, food engineers have to add a lot of ingredients that don't necessarily

---

10. On their Web site, they list these things in the same set of bullet points, which perhaps gives you some insight into the mind of your typical food engineer.

11. We showed this nota bene to Dr. Lipton, who told us, "In all fairness, we have since called that paper 'our greatest crime against the culinary arts,' and have put chefs in the driver's seat to repent."

12. Or maybe not. Dr. Lipton, sounding like the host of a really weird infomercial noted, "You would be surprised how much of your food is already extruded."

help the flavor. For the foreseeable future, you're probably better off at the muffin store than the bio lab.

Plus, as far as we can tell, you really just can't trust food engineers. I mean, they're currently trying to "close the loop" on astronauts. Imagine what they'd do to you if they could.

7.

Augmented Reality

An Alternative to Fixing Reality

Your boss comes to your cubicle to yell at you about something. You know you can't stop him, so you settle in to half listen for ten minutes. Then you remember you have a tiny computer screen embedded in your contact lens.

You wink. Your boss is confused for a moment, but then continues ranting. Before your eyes, the world changes. Palm trees sprout in the background. The light dims ever so gently. A delicate pink songbird alights on your stupid boss's stupid comb-over.

A molecule generator in your nose releases the scent of ocean breeze, and the pair of tiny speakers in your ears makes the sound of crashing waves. Your right-ear speaker makes the sound slightly sooner than your left, and you look over your right shoulder to see the blue Pacific.

The processor on your desk detects the voice of your boss and converts whatever he's saying to a sports update, which you listen to at ease while the sea breeze laps against the palms. Just as your boss, in real life, asks if you're planning to actually do your job for once, a sportscaster says that your favorite team lost again. "NO!" you shout. "NO! WHY?!"

Fortunately, when you get home to your (apparently) gigantic 200-square-foot mansion, your virtual spouse doesn't judge you for getting fired from fourteen jobs this year. You put a thin polymer coating over your tongue, pull some soy protein from the pantry, and decide it will taste like Kobe steak tonight.

This is what reality could be like, with a little augmentation.

The big idea with augmented reality (AR) is that you can take the real world and overlay virtual elements onto it. It's sort of like adding a little magic to the universe. This is different from virtual reality (VR) because virtual reality blocks out all *real reality*. One way to think about it is to visualize yourself as a brain connected to a bunch of sensors for things like taste, touch, sight, motion, balance, and so on. This is how AR researchers view you, so you might as well do it too.

All these sensors are constantly taking in information from the environment around you. In a complete VR system, those sensors are 100% occupied by fake inputs created by a computer. So you're standing in a tiny room, but your senses tell you you've been transported back to the Cretaceous period and a *Tyrannosaurus rex* is heading your way. In an AR system, those sensors are only partially occupied by fake inputs. So you're standing in the middle of a real mall, and a *T. rex* is preparing to head your way as soon as he finishes his soft pretzel.

Right now, we mostly adjust the visual sensors (sometimes called "eyes"), for reasons we'll get to later. Adding virtual objects and information to your visual sense alone has all sorts of applications. In any situation in which you want to still interact with reality, but also get more information (like combat, surgery, or construction), AR might be very useful indeed. If perfected, it could mean better outcomes in fields like these, with less training necessary. In other words, cheaper, better services. Also, you can lie to yourself about your life more effectively than ever.

With the recent release of Pokémon GO, AR has suddenly become commonplace. We won't delve into it too much, since you're probably playing it right now as these printed words sit idly on a shelf. But we see Pokémon GO as an early step in a technology that has applications beyond gaming.

Your authors are in a group we call the Saddest Generation—those young enough to know about Pokémon, but old enough to be confused and bewildered by its popularity. That said, AR promises to make fantasies of all sorts come true. And no, they aren't all sexy dreams. Some are about revenge or getting rich.

There are a number of ways you might achieve this AR, and given the number of senses you have, a perfect AR system will be quite complex. At the present moment, the most common method is to use some means (right now, usually a tablet or phone) to project an image into your eye that is "in registration" with reality.

Registration just means that the virtual stuff is cooperating with the real stuff. For instance, if you have an AR bunny running around the room, you don't want it to run through things. Or if it does run through things, you want it to look like it got hurt.

This is a lot more complex than it might sound at first. Imagine you've got a headset that projects AR into your eye. You look at a table, and the headset projects a letter from your imaginary spouse, Brad Pitt, saying, "I love you." Sure, you programmed your AR to leave this letter for you today, but just like when a spouse remembers your birthday after you reminded them the night before, you still appreciate that AR Brad Pitt left a note.

When you turn your head left, Brad's letter needs to go right. It also needs to change its apparent angle toward you in order to correspond to the actual table. If it is even slightly off, it will look weird, the fidelity will be lost, and you'll remember that what you actually have on the table is a restraining order.

Achieving all this is an extremely difficult challenge, requiring great hardware, great software, and great understanding of how human vision and cognition work.

Hardware is advancing nicely. The earliest machine you could argue was augmented reality was created by Morton Heilig in 1962, and called the Sensorama. This machine was purely mechanical, meaning there wasn't a programmable computer involved. Whereas a modern device would have a program that knew to release the smell of trees when you went through a virtual orchard, this device had smells that were simply triggered five minutes after a video started, corresponding to when the video was showing an orchard.

The Sensorama played videos through little eye portals, while generating wind, playing sound, vibrating, and releasing chemical scents, giving you, for instance, the sensation of riding a motorcycle. Why you wouldn't just buy the damn motorcycle is anyone's guess, but Heilig was looking toward the future. In the patent he filed, he discussed applications for the military, entertainment, industry, and education, which are the exact topics most frequently discussed as applications for AR research today.

Some of you may remember that so-called virtual reality headsets looked like they would finally break out in the 1990s, as computers and monitors got cheaper. If it didn't happen then, why should it happen now, or in the future? Well, those '90s systems had problems. They were really, really expensive, and the simulations were quite bad, and well . . . they had a tendency to make you puke. Puking is actually a general problem with AR and VR systems. One current theory of motion sickness suggests that when you don't *feel* like you're moving, but you do *see* that you're moving, your brain decides that you've been poisoned. Hence your desire to run to the nearest bathroom.

The motion sickness problem is especially acute if the machine creating your fake reality is a consumer computer from 1993. If you're wearing a headset and you turn your head, but reality doesn't turn for a whole extra second, your eyes may not believe this is real, but your stomach sure does.

Like most '90s trends, VR was wisely shelved. This represented a setback for people interested in AR, who hoped that popular virtual reality headsets would make the relevant technologies cheaper.

But, just as Kelly has done with most of the so-called music of the '90s, a few scientists have persevered in AR, hoping for a resurgence. And, unlike Kelly, they have good reasons to be hopeful.

Today, pretty much everyone wears a computer in the form of a phone. In fact, it's a bit better than that. A typical smartphone doesn't just have a computer—it can take pictures and video, detect its own orientation, detect gravity, determine its position on the face of the Earth, and other neat things. These detection abilities are especially useful when you're trying to overlay a fake reality atop real reality. In addition, the smartphone isn't isolated like a computer from the '90s. It can talk to other computers that have way more memory and processing.

Somewhat by coincidence, today you are walking around with a lot of the equipment an AR researcher would want you to wear. The trick then becomes getting the software right.

One early way to add virtual objects to actual life was the "fiducial marker." Basically, a fiducial marker is an object placed in reality that is easy for a computer to recognize visually. Visualize something like the now common QR code. Imagine you have a table with a QR code in its middle. For simplicity, let's imagine you are wearing an augmented reality headset that projects images into your eye. The headset's cameras see the QR code and determine two things: (1) that its pattern codes for "put a vase here," and (2) that you're looking at the QR code from a particular angle.

As you move, the headset detects the changing orientation of the QR code and adjusts the vase accordingly. If it works right, you perceive a vase sitting on your table, even if you walk around or jump up and down. In other words, the fiducial marker serves as a simple bridge between augmented reality and actual reality.

Current AR research has moved beyond traditional fiducial markers. In fact, perhaps as a sign of rapid progress, we were told the term "fiducial marker" was no longer cool, despite it being used in texts we found from just a few years ago.

These days, the programs have become smart enough to figure out where to place objects on their own, though markers of various kinds are still handy because they can give the computer a lot of information quickly.

But markers have problems. For one, they can get blocked visually. This isn't a problem for a real vase because . . . well . . . the vase is still there even if you're not looking. But imagine you dip below the table and look up. At your angle, you should be able to see the vase, but your headset can't see the marker. Your headset decides the vase no longer exists. This is bad because, generally speaking, most of us would like reality to keep existing when nobody's looking.

You can solve this problem by having an extra camera or by having additional markers that orient your headset, but the more stuff you need, the more cumbersome the AR experience becomes. Making the interaction between real and virtual worlds subtle is crucial to making the experience work.

One neat idea in AR is that you could be walking through a forest and have your experience enhanced with knowledge about local ecology and history. For instance, if you pass a live oak, perhaps the computer tells you what a live oak is. As you come closer, you see it has a fern growing on it and is infected by some gall wasps. Your headset displays information about these things too. It also tells you that, by the way, a Civil War battle happened in this forest in 1864, and offers you the option to see a virtual reenactment laid over.

This is all awesome, but it would ruin the mood (and be hard to set up) if you had to put a QR code on every object the user might be interested in.

So a big area of current research is how to use regular ambient markers to determine all the stuff a QR code might tell you. That way, instead of putting markers all over the Eiffel Tower, you'd have a device that just recognizes the Eiffel Tower.

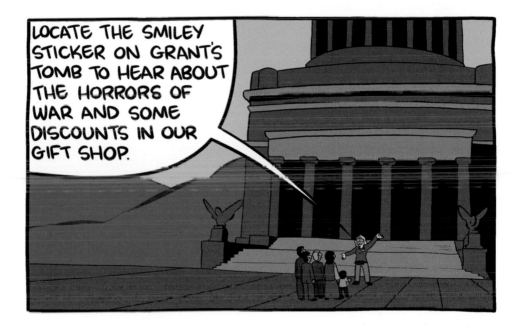

"*But wait,*" you say. "I'll just use my GPS. My GPS knows where the Eiffel Tower is." *Nope.* Won't work. GPS only tells where you are on the surface of the Earth, and it's only accurate within about a meter. Good AR requires more accuracy than that. And GPS accuracy is even worse when you're measuring elevation. So you try to activate a virtual tour guide in Edinburgh, but your virtual tour guide appears 10 feet overhead. You look up, and suddenly wish he weren't in traditional Scottish dress.

But GPS is a good starting point. It might be able to tell your computer that you're in a certain park or near a certain lake. And we know the computer *should* be able to tell exactly where you are from visual cues. After all, this is basically what humans do to orient ourselves. If you're lost in the woods, you try to find a tree with a distinct appearance, or to see in what direction a large distant landmark is. In principle a computer can do the same thing. But although a human can fairly quickly determine what is "distinct," it's hard to explain distinctiveness to a machine.

One approach goes like this: Imagine you're walking near the Empire State Building. Your computer knows roughly where you are, but in order to project a giant gorilla onto the building, it needs to know your exact spatial coordinates and exactly where you're looking. So in order to map your field of vision, it takes a photo. It now divides that photo into sections. It decides which are interesting and which aren't by looking for variations in intensity.

For instance, if the computer gets a square that's all sky, there is probably no variation. This tells the computer it's not an "interesting" point. On the other hand, if it gets a part of a window, there'll be lights and darks and geometric shapes. This *might* be interesting.

The computer then compares that image to a database of other images, which (thanks to GPS) are known to be of nearby landmarks. By doing this repeatedly as you move around, eventually it gets a lock on your exact position, as well as the orientation of its onboard camera. This might sound like a weirdly artificial method, but it's more or less what you do to figure out where you are as well. When you're lost, you don't reference the sky at day for position because a big blue heaven doesn't provide much specific information about where you are. You *could* reference the sky at night, because the stars and their position in the sky carry information about your location. You don't reference the ground when trying to orient yourself, but you *do* reference your distance to a distinct building. You're so used to your frequent haunts that you do all of this automatically, but try to remember how your brain worked things out the last time you went back to your hometown after being gone for a long time.

Part of the problem with this approach is that it requires a lot of computation and a huge amount of reference images to compare with your "interesting" pictures. So scientists are working on other approaches to simplify the process.

Dr. Caitlin Fisher of the Augmented Reality Lab at York University suggested that there might be an easier way to go, at least for some applications. "Artists," she says, "use mixed reality that sidesteps the registration problem. You can have images that might just float or be in the sky or tiny images on the ground. The registration problem is an urgent one, but it's not necessary to overcome it for all good augmented reality experiences. . . . I think one of the reasons you find a lot of the early projects involve ghosts or spirits is because you don't have to have them perfectly registered in the real world but you can still have this very cool experience. One of the reasons why your thing can be floating, [and] people don't necessarily say, 'Oh, poor engineering,' they say, 'Wow, it's a ghost.'" We're not quite sold on being virtually surrounded by ghosts, but it might be something fun to try out on unsuspecting children.

This sort of thing is fine for more artistic projects, but, for instance, if you're using an AR application to do surgery, you don't want virtual incision markers to stray too far from home.

One way to get better location detection without making your computer cry is to have better sensors. A technology called LiDAR (light + radar = LiDAR) bounces laser light off objects, then analyzes the reflection. LiDAR can generate accurate 3D models of the environment, which is exactly what you'd want for augmented reality. Instead of comparing a universe of flawed 2D image files, you get the outline of local buildings and compare to a single 3D file. Sounds great! The problem is that it's historically been superexpensive. Like, only used by huge government agencies expensive.

But over time the cost has come down. In fact, one of the reasons autonomous cars are starting to come to market is that you can get a decent LiDAR system on your van for only a few thousand bucks. The downside is that the lightest ones still weigh around 10 to 20 pounds.

Still, the technology for visual AR is coming along nicely. "But," you interject, "what about my other senses? I WANT IT TO SOUND LIKE BIRD-SONG WHEN I TURN THIS PAGE!"

Well, first of all, stop yelling. It's puerile. Second, well, the majority of current research is about visual technology. Humans are *really* into looking at stuff.

But a few scientists and engineers are working on audio, smell, and touch technologies.

Audio is relatively simple compared to video, but it does present its own difficulties. Imagine you want to simulate the sound of a car going by. There are three major issues: (1) The sound must hit one ear earlier than the other, giving a sense of the car's location, (2) the sound must change pitch and intensity, giving you a sense of the car's motion, and (3) this is where it gets really tricky, the sound of the car should (literally) reflect its environment. So if you're in a canyon, the car sounds should echo. If you're on a field, it should not. This takes us back to a lot of the issues discussed earlier for video: To get better AR, you need more information and computation.

Smell is really tough. With light and sound, there are simple elementary components. To produce any color, you just need various wavelengths of light. To produce any sound, you just need the right volume and waveform. Smells are far more diverse and complex. Even if you had a huge set of preloaded smells, it's not always the case that you can mix them to make new stuff. For instance, "apple smell" plus "pie crust smell" doesn't exactly equal "apple pie" smell.

In principle, you could have a machine that could custom manufacture molecules on the fly and spritz them into the air. There are, in fact, machines that can generate made-to-order molecules. But currently this is an extremely expensive process, used for industrial purposes. And while visual and perhaps audio AR might have lots of clear applications, it's not obvious what utility virtual smells would provide.

The sense of touch is a bit more active as a research area, but progress here is still quite limited. Much of the current research is about having a "haptic pen" you can use to "touch" virtual objects. The basic idea is that you use a headset to see a virtual thing, and then you have a pen connected to a computer that knows where the virtual thing is. If you try to poke through the virtual object with the pen, the machine won't let you. If you drag the pen along the surface, it wobbles so the fingers gripping the pen get a sensation that's as if they had actually dragged a pen along a textured surface. Okay, so it's not quite as good as getting to actually have touch feedback for Virtual Punch-Hitler-in-the-Mouth, but stabbing with a pen is a close second. Plus, it has some real applications, such as training for surgery, or digital sculpting.

Where Are We Now?

A lot of the sources we read on this topic were written between 2010 and 2014. The later ones tend to be really excited about how Google Glass is gonna change everything. Whoops.

Google Glass is generally considered to have failed because when people see you wearing it, they want to punch you in the face. No, really. For instance, in 2013 the CEO of Meetup.com literally said to *Business Insider* reporters,

"Google Glass? I'm definitely gonna punch someone in the face wearing Google Glasses."

So one feature you'd want in the future of AR is a display that doesn't get you punched in the face by tech millionaires. The trick there may be miniaturization.

Innovega is one company working on an AR contact lens. It's not quite to the point where a contact lens can do the whole job, though. In fact, you have to wear a special pair of glasses over the contact lens. On the plus side, unlike Google Glass, it pretty much actually looks like a pair of glasses.

In general, sensors and computation systems are getting cheaper, faster, and smaller. This provides fertile ground for experimental projects, of which there are many.

Dr. Mark Billinghurst, currently at the University of Canterbury, came up with a concept called a "magic book." The idea is that the images on a book work with an AR machine to enhance the experience. One group at the University of Nebraska created a "magic book" called the Ethnobotany Study Book and had little black-and-white drawings of local plants. If you were looking at a drawing through an AR machine, you would see a virtual version of the plant emerging from the page.[1]

Dr. Jonathan Ventura of the University of Colorado created an augmented reality program while he was a PhD student at the University of California, Santa Barbara. It was a program meant to be run outdoors, and it allowed you to add things like spaceships and trees to your perception of the UCSB campus. According to Dr. Ventura, "I set up a system where I would map out, for example, the UCSB campus with lots of images and basically create a 3D model of the campus. Then, I could take out an iPad and just hold up the iPad and it would take a picture of the surroundings, match that to its model, and then figure out exactly where the device is."

We consider the phrase "things like spaceships or trees" to mean "space-

---

1. If you're curious what this would look like well guess what? *You're reading a magic book RIGHT NOW.* Go to http://www.SoonishBook.com to download a free app that will cause a familiar structure to virtually grow out of your cover.

ships." Dr. Ventura is a bit more pragmatic than us. He points out that programs like these could be very useful tools for landscape design architects, who could virtually landscape an environment and show it to a client before actually beginning construction.

Dr. Gerhard Schall at Graz University of Technology created an AR system in which city workers can get "X-ray vision" of city infrastructure. For instance, they can look down at the street and see the electrical and plumbing systems underground. Systems like this have a lot of potential applications, not just for maintenance but also for disaster relief.

By having a virtual projection of how things are supposed to be, a relief worker should be able to assess damage more quickly. For example, one metric for deciding if a building is badly damaged is "interstory drift." That is, on a scale from Empire State Building to Tower of Pisa, how screwed are we? Determining how much a building leans is actually somewhat difficult, especially in the aftermath of an earthquake, when equipment is scarce and time to assess each building is limited. An AR system proposed by Dr. Suyang Dong could project what the building *should* look like over what the building *does* look like, allowing inspectors to make rapid accurate judgment.

One idea that a number of groups have worked on is a "virtual mirror interface." You know how as a kid you had this fear that you'd look in the mirror and someone else would be *inside it*? What if we could have that for real?! The idea has been around since the 1990s as a relatively simple version of AR, since it only has to project via the mirror and it only has to know the layout of your house. The idea, other than just generally being neat, is that you could have a virtual helper or pet who "lives" in the mirror. So the mirror acts as a sort of window into an augmented reality. Personally, we find the idea slightly terrifying. Like, there is just no world where you pass your mirror on the way to the bathroom at night and a guy is sitting on your couch in the reflection and you think, "This is fine."

In fairness, the helpers will probably be adorable. At least, right up until they betray us in the Robot Uprising of 2027.

The Fisher Lab (remember the lady with the ghosts?) explores the use of AR in storytelling. For example, how can AR be used to make films you can walk through? Or how can AR be used to bring historical people and places to life? Imagine walking through locations along the Underground Railroad, seeing virtual actors tell stories of their captivity and flight. Imagine visiting the Battle of Belleau Wood site, and seeing the 1918 offensive before your eyes. Or imagine visiting the Colosseum, seeing the grimace on a gladiator's face as Nero shows no mercy.

Dr. Fisher also throws the best children's birthday parties in history, which include Harry Potter–style AR sorting hats, fairies soaring over stones in the backyard, and fairy dust that would settle into the children's hands. She notes that the children's imagination probably didn't require augmentation though: "I don't actually think small kids need augmented reality as much as we need augmented reality to think more like small kids again. I think that there are these amazing, joyful practices we could have. On that small level I think individually it could make our lives awesome."

A few people have tried using augmented reality for therapeutic reasons. Our favorite experimental therapy was by Dr. Cristina Botella, who had the idea that you could cure phobias using AR. See, one of the best ways to get over an irrational fear is repeated exposure to it. But there's this problem. When you take someone who's afraid of roaches and make her repeatedly get in a box filled with roaches, she might decide to find a new psychiatrist. Or, as the scientists say, you'll have a "high rate of attrition." Dr. Botella wondered if you could work out a compromise, where you simply project hordes of horrifying insects into the subject's eyes. The study we read only had six participants, but all six of them seemed to come away from the experience with lessened phobias that were maintained over time. Of course, maybe they just said that to make Dr. Botella stop.

Finally, we're particularly excited about DAQRI's Smart Helmet. The innovators at DAQRI made an interesting observation—hard hats can be modified to include AR without fundamentally changing anything about hard hats. The sensor and computers are embedded in the hat and the visuals are displayed on the eye shield. Maybe Google Glass should take a cue from DAQRI: If you wear a work helmet along with your computer, you don't look like an asshole. Plus, if the CEO of Meetup.com tries to punch you in the face, well, you've got some protection.

These Smart Helmets have the potential to make us much more efficient and perhaps save a lot of lives. We spoke with Gaia Dempsey at DAQRI, and she told us about a recent study done by DAQRI, Boeing, and Iowa State University, comparing training with AR to traditional training. "We compared paper instructions to augmented reality instruction for a complex assembly task. This particular assembly task was a 50-plus [step] process of assembling [an] aircraft wing tip for first-time trainees. The group that had augmented reality instructions was able to reduce their job completion time by 30% and reduce their error rates by 94% on their first try. On the second try, they got their error down to zero, which is huge." Your authors have a strong preference for zero errors in any process related to aircraft assembly.

Concerns

Perhaps it's occurred to you by now that having every single human equipped with constantly active sensor arrays communicating with centralized servers might, maybe, raise some privacy issues.

We read about a piece of software called Recognizr. The software can detect the features of people, turn their faces into 3D models, then recognize them later. For now, Recognizr is opt in. The idea with lots of this sort of software is to bring social media into real life. There is some good potential here—imagine you go to work and a little display above your coworkers' heads gives you information, like whether it's their birthday, or they just got back from a trip. The downside is that all the regular social media privacy concerns come into real life too. If a lot of people have face-tracking software, however dispersed the data is, a bad actor could reconstruct everywhere you went during a day, and perhaps even make good guesses at your emotional state.

It gets worse. Remember, the ultimate AR machine doesn't just track and store visual data. It takes 3D scans of everything. It smells. It hears. A lot of modern commerce relies on companies having access to a huge amount of data. This is why Amazon and Google can tell you what you want before you know it. But how do we feel about a world where a thermal camera detects that you're a bit hot and sweaty and pops up an ad for Starbucks' New Berry Blossom Iced Whatever? Or maybe a face scanner noticed you grimacing a lot today, and recommends you "Ask your doctor about Zoloft"?

Even the good qualities of social media carry weird consequences when ported into real life. With AR, your boss "knows" it's your birthday, or that your husband just left you, or what your favorite TV show is. He knows all this because he has subtle AR glasses, which project data from your online profiles over your head. This is a concern for two reasons: First, in some sense it's redefining what it means to "know" something. By now, we're all used to the idea that so-and-so remembers your birthday because Facebook told her to, but what if it's more than just birthdays? There's something off-putting about a world where a lot of someone's working knowledge of you is externalized and projected onto a heads-up display. And if you *don't* have AR glasses, there's a tremendous information asymmetry between you and the people who *do* have them.

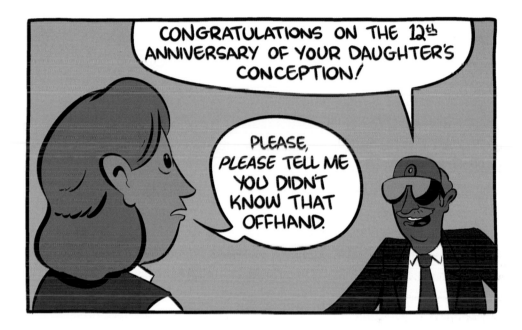

Since information asymmetry is a serious matter in warfare (especially so in modern warfare), one idea is to give this sort of software to peacekeepers. Imagine if a soldier's job is to police a village. It'd help if he had glasses that could recognize faces and display stats on his heads-up display. He's now a soldier who remembers every villager's name and knows every villager's needs, not to mention his religion, politics, and friendship networks. Does this make the soldier more or less empathetic? And regardless of the effect on the soldier, how do the villagers feel?

One topic we came across repeatedly in our research is the notion of "diminished reality." The idea is that there are times when you'd like *less* real sensory input. In a sense, total immersion VR does just that—it completely blocks out *real* reality. AR is more of a mix. In principle, you should be able to turn a knob between total reality and total virtuality. There are certainly good uses for this sort of thing—people with anxiety disorders or disorders arising from trauma may wish to selectively deaden certain types of sense experience. But diminished reality may take on an ominous flavor in contexts where hard choices must be made. For instance, could you block out homeless people on your way to work? Or could a soldier in a combat zone block out the emotion on the faces of enemy fighters?

A fundamental notion with AR is that its users must have access to a tremendous amount of data about the world. This is especially true for extremely advanced ARs, in which all sorts of modifications are made to the world. Consider the simple case of an AR experience in which all squirrels are constantly shooting bees out of their eyes. You need more than a computer capable of recognizing squirrels and tracking their motion relative to yours; to really complete the illusion, you should now and then see bees firing out from, say, behind a tree or behind a fence. You'll need really good sensors to do this, or you'll need there to be cameras of some sort all over the place. In other words, to get a more immersive experience in an AR world, you need more data about reality. If AR became popular, it might lead to a consumer push for ever less privacy. And regardless of how consumers feel, there will be far more data in the hands of corporations and governments.

Dr. Alan Craig of the University of Illinois at Urbana-Champaign pointed out another concern: Who gets to control what gets projected where? For example, suppose you own a store. Someone in a highly popular augmented reality system graffitis "The owner of this store is a fool and a cad" on your wall. For the sake of argument, let's say it's not true. Do you have a right to get it "removed" from the augmented world? After all, it's not *touching* anything you own.

In fact, while we were finishing this book, Dr. Craig's concern went from hypothetical to real. In Pokémon GO, a "PokeStop" (or a place where you can stop to get free items for the game) ended up in a Holocaust Museum, and the museum had to request that players please stop playing the game in a museum meant to honor victims of the Holocaust.

Later, people were literally looking for Pokémon at Auschwitz. At first we wondered if this was a narrow oversight pertaining to Holocaust memorials. Then we saw an article in the *Telegraph* titled "HIROSHIMA ANGER OVER POKÉMON AT ATOM BOMB MEMORIAL PARK." So . . . at least they're equal opportunity offenders.

Dr. Craig noted that AR can get hacked. So, if and when Pokémon GO fixes this issue where Squirtle is violating the sanctity of the dead, bad actors could still create problems. They might also create danger. What if you're wearing your AR glasses while driving, and some hacker sends a virtual pterodactyl careening toward your car, causing you to reflexively swerve? It'd be awesome, sure, but also you'd probably be dead. If AR is ubiquitous, your perception of reality is hackable. And so are the perceptions of people and groups around you.

Finally, does there come a point where we lose track of what is artificial and what is real? Dr. Craig said that movies are currently a mix of real and computer-generated images, and are often done so well that we can't differentiate reality from illusion. Could AR get so good that we may walk around one day without knowing which aspects of our environment are part of the real world and which are only projections? Does it matter?

How It Would Change the World

Of course, the most popular idea is that AR will revolutionize entertainment. Being able to bring magic into reality presents artists with an entirely new canvas. In principle, infinite new worlds of ever-increasing complexity could be rolled into this world, but it is difficult to know how a new technology will shape the future of art, especially as AR becomes a more serious medium.

One exciting possibility is the improvement of education. AR technology allows people to interact with conceptual entities, which might be especially helpful in fields where the subject matter is hard to visualize. Three-dimensional concepts in physics often baffle students, but they might learn more quickly if they could push those concepts around, visualize, or perhaps even touch them. Imagine a chemistry class where the atomic interactions are displayed as a virtual object in front of your face.

Dr. Craig worked with the University of Illinois's Veterinary School to create a life-sized fiberglass cow. If you view the cow through a smartphone using an app, you can see where the internal organs of the cow are located. This is an amazing tool for veterinary students, who can visualize the locations of organs in three dimensions.

Over human history, we have slowly off-loaded work from our own brains. Writing allowed us to stop memorizing everything. Filing systems made it so we didn't have to remember the location of information. Modern search systems mean that a lot of the work of researching a topic is simplified. Successful AR might mean that a whole constellation of mental activities can be off-loaded to the machine.

For example, when you need to fix your own printer, you probably find an online guide and work your way through it until you understand what you're doing. An AR system might simply show you step-by-step instructions as you go. By this means, you accomplish the task more quickly and successfully. A similar approach might be taken to cooking or building things.

This may seem small, but it could represent a huge efficiency gain in many areas. Workers could be trained faster and avoid more on-the-job danger. Conceivably, an AR machine could even make an expert better at her job. For instance, an AR helmet might alert the worker to a subtly dangerous structural change that's easy to miss.

AR may become common in medicine as well. Anatomy apps like Complete Anatomy Lab's upcoming Project Esper could be a useful tool for medical students. There are also more serious uses. One company called Illusio made a program that allows virtual breast augmentation. Women interested in breast surgery can see a virtual "reflection" of their breasts. The virtual breasts could be adjusted along metrics such as perkiness and amount of cleavage. Perhaps they should consider selling a home version.

Joking aside, there's real utility here. If you need cosmetic surgery, getting a shared vision between yourself and your doctor might spare you a lot of problems.

Other apps may even be used during active surgery. If a surgeon could project your MRI scan onto your body while she does surgery, then she may be able to make smaller and more precise cuts. She may also be able to project a person's former appearance onto her body following a disfiguring accident, and this projection may allow for more accurate reconstruction surgeries.

Emergencies in wartime or in rural areas of poor countries may call for immediate specialized surgeries for which the nearby surgeons have not been specifically trained. AR may be able to help with that. Programs are being created where surgeons who specialize in the needed surgery can remotely guide a distant surgeon through the procedure. They can even "draw" on the surgeon's field of view if the surgeon is looking through an appropriate screen. While the person under the knife would probably prefer if the specialist were in the room, the next best thing when time is short may be having the specialist drawing a big arrow for the acting surgeon, saying "Cut here!"

In general, AR might allow us to quickly acquire skills that formerly required a huge amount of training. This could dramatically increase efficiency, and may save lives in situations where the job being done is dangerous if mistakes are made.

At its grandest, AR offers us the opportunity to remake the world the way it is in our imagination. As humanity has moved from the savannah to the skyscraper, we have discarded many myths and daydreams that, though false, were comforting. But with the right technology, we can put dragons in the sky or have pixies dance in the garden. We can wander through the imaginary worlds of loved ones and explore facets of our own personalities. We can even, in some sense, make the dead rise again.

So, you know, let's not hate on Google Glass too hard, okay?

Nota Bene on the Nasal Cycle

So Zach was doing some reading on augmented reality, and he noticed something. Humans have two eyes and two ears, which in both cases help us determine the position of things around us. Having two eyes at a slight distance from each other allows you to see things from two slightly different angles. Your brain combines the two visuals to give you a more three-dimensional image of the world, which helps you figure out how far in front of you an object is. And having two ears helps you figure out where a noise is coming from. If a noise sounds louder in your right ear than your left ear, then the noise is probably originating from somewhere on your right side. But there's one other double thing on your face—the nostril. We wondered: Does having two nostrils rather than one help you locate where a smell is coming from?

As good researchers, we went to Twitter first. Twitter said we were wrong and dumb. We were now very motivated to be right, whether the facts were with us or not.

Long story short, we were pretty much wrong but maybe kinda right for some animals. It turns out a few animals, like dogs, *can* collect separate air samples from each nostril, and may be able to compare those samples to figure out where a smell is coming from. Also, Kelly knew of one paper that suggested something *like* depth perception via smell in snakes. The snake sticks out its forked tongue to gather chemicals in its saliva. It then wipes its tongue over two depressions in the top of its mouth. The saliva then gets sucked into little bulbs known as vomeronasal organs, where the chemicals in the saliva are processed.

Lots of animals, such as mice, dogs, and goats, have this organ. But only snakes and lizards have the forked tongue. The author of this paper, Dr. Kurt Schwenk (whose last name would make a great verb for sampling odors with a forked tongue), showed that the forked tip allows its owner to detect two points on an odor trail and compare them to determine which way to go. It is also possible, but not yet confirmed, that snakes can use their tongues to sample two points in the air to tell what direction an odor is coming from.

You probably don't notice, because it's automatic, but when you want to figure out where a good smell is coming from, you position your head in one place, sniff, then position your head somewhere else and sniff again. This is because you have a boring unforked tongue and are thus unable to schwenk.

In our vain attempt to find smell-depth perception in humans, we stumbled upon the amazing literature on the nasal cycle. You already know about it intuitively, but probably haven't considered it—at any given time, most of your breathing is through one "active" nostril. Your nostrils alternate which of them is active throughout the day, typically cycling every two to eight hours. So why two? Why not one big *mega*-nostril?

Let's talk about your mucociliary escalator. Yes, you have an escalator in your body, whose primary passenger is mucus. It is "mucociliary" in that cilia (imagine little bristles) in your body do the escalating. In particular, they move the mucus toward your mouth so you can swallow it or hock it up.[2] This keeps your nose clear for breathing and smelling, and it keeps your lower respiratory system relatively sterile.

But if you're using just one nostril for most of your breathing most of the time, eventually your mucus escalator will dry out. Once dry, the elevator doesn't work so well, and the sensitive nasal tissue can get hurt. The solution? *Alternating nostrils.*[3]

At any given time, one of your nostrils is more ready to permit air than the other. This is achieved by your "nasal venous sinusoids," which become engorged one at a time in a process called "the nasal cycle." We use the term "engorged" advisedly. One paper[4] refers to the nasal venous sinusoids as "a spongy tissue, similar to erectile tissue."[5] Cycling your nostrils means that one of them is always getting a break from what Dr. White calls "air conditioning or house-cleaning duties."

At least one author proposes that the nasal cycle is upregulated as a defense against disease. Essentially, your inside-the-nose-erectile-tissue is working overtime to make fluids to protect you against more bad stuff in your environment.

---

2. According to Dr. David White, who was kind enough to read through our mucus stuff, "airways produce around two to three liters of mucus per twenty-four hours. Most of this is swallowed and serves to assist the mucus lining in the gut."
3. There is a lot of serious science on this. One paper (White et al., *BioMedical Engineering OnLine* 2015, 14:38) was a complex mathematical model of the nose, which used MRI scans. And we quote: "This model considers the two complex nasal cavities as a series (k) of aligned tubes of varying hydraulic diameter. . . ."
4. Eccles, *European Respiratory Journal* 1996, 9:371–76.
5. If any of you are reading this with a clogged nose, we apologize.

We got pretty thick into this nose business, but our favorite find was the apparently extensive literature on the cognitive effects of Unilateral Forced Nostril Breathing (UFNB). According to Dr. White, "Think of this as forcing the nasal cycle so that airflow is dominant on a side of choice. This influences the hypothalamus, which controls many other physical and neurological systems."

The bottom line is this: Many experiments have been conducted in which undergraduates were forced to breathe through a certain nostril, then made to take exams. There is apparently decent evidence that the nostril you're currently breathing through may affect your performance on intelligence tests and emotional tests. A number of papers also found associations between which nostril was currently active and the onset of hallucinations and schizophrenic episodes.

We remain a bit skeptical but entirely satisfied as we imagine what was done to those starry-eyed young college students.

8.

Synthetic Biology

*Kind of Like Frankenstein, Except the
Monster Spends the Whole Book Dutifully
Making Medicine and Industrial Inputs*

We humans have been tinkering with biology for a long time. In fact, it's kind of our thing.

We've been genetically altering biology, including the foods we eat, for at least 10,000 years. If you look at our primate cousins, their food tends to be seedy and high in fiber whereas our favorite foods are things like cake, beer, and beer cake[1]—no-fuss calorie conveyances.

We've gotten pretty damn good at altering biology. One time, we took a single species called *Brassica oleracea* and turned it into every vegetable you hated as a kid—brussels sprouts, cauliflower, broccoli, cabbage, kale, kohlrabi, collard greens. YES. All one species, slowly modified over generations into a thousand okay-tasting forms, each more cheese-requiring than the last.

---

1. Yes, it exists. The word either denotes a cake in which beer is used to add flavor and fluffiness, or (apparently) a bunch of beers arranged neatly in stacked cylinders, sometimes accompanied by fireworks, whiskey, and lottery tickets. God bless America.

We do this to animals too. Remember when we took the noble wolf—spirit of the forest and tundra—and converted it into a shivering, bug-eyed rat, dependent on the nearest blond socialite for diet kibble and a pink sweater? There is no greater expression of man's dominion over nature.

All these changes occurred because humans started controlling the breeding of the life around them to have more qualities we like and fewer qualities we dislike. When they did this, they were unknowingly altering these species' DNA. But our ancestors were altering DNA incredibly slowly, making small changes over many generations, and tinkering with traits that already existed.

Once we understood DNA better, we began trying to manipulate it. For example, there's atomic gardening. The basic idea is that you get a radioactive substance, then grow your garden in a circle around it. Now you're mutating all the surrounding plants at an increased rate, increasing the chance that you'll find something really cool. To be clear, irradiating plants does not result in radioactive descendants. The radiation is just a convenient method to modify a plant's genes, perhaps causing it to transmit those mutant genes to its offspring.

This may sound bizarre, but the (mildly) euphemized field of "mutation breeding" has created a lot of your favorite foods, like modern "Ruby Red" grapefruit and "Golden Promise" barley (which you've probably consumed if you've had a few Irish beers or whiskeys).

But this method is still quite blunt. You change a bunch of DNA and hope that, by dumb luck, there's a noticeable useful mutation. But what if we could take an entirely new approach—instead of selecting good traits or making random changes to an organism's genetic code, what if we could precisely alter genes, knowing exactly what the result would be? Sure, we could make a better grapefruit, but we could also take life beyond its current boundaries. We could convince bacteria to be little factories for medical chemicals. We could get microbes to take readings in areas that are hard for us to probe. We might even be able to alter human DNA, perhaps even in currently living humans.

We are going beyond the realm of natural biology, into what is called synthetic biology.

To understand how all this might happen, you need to know a little about DNA. Let's do a quick rundown.

DNA

In all multicellular organisms (like mushrooms and humans), cells have a distinct inner portion called a nucleus. Inside the nucleus are really long molecules called DNA. You can think of DNA as an especially lengthy rope ladder that twists around and around in a corkscrew shape. This is the famous "double helix."

The "rungs" of the ladder are made of two small molecules (one from each side) that fit into each other, hand in glove. Well, maybe we should say hand in glove *or* foot in shoe, since there are two ways these pairings happen. These small molecules are called bases, and they come in four types, abbreviated T, A, C, and G. Base T always pairs with A (hand in glove), and C always pairs with G (foot in shoe).

The result is that if you pulled apart this spiral ladder, yanking the hands and feet out of the gloves and shoes, you would see a long string of bases on each side. If you read them in order from beginning to end, it'd go something like "AAGCTAACTACACGTTACTG" only much, much longer. Like, one hundred and fifty million times longer in humans. These letters encode most of the information your body needs to do all the stuff you do.

So what the hell does that mean? Most of the stuff going on in your body is done by proteins. People generally think of proteins as the stuff you eat when you take a bite of chicken, but the word "protein" refers to an enormous category of molecules that function as the little machines that do pretty much all the tasks in your body. DNA is, so to speak, the library for how to make proteins.[2]

Now visualize the DNA ladder opening up, with each rung splitting down the middle, so that it's a long string of T, A, G, and C bases. On this newly open surface forms a new molecule called RNA. RNA is a sort of mirror molecule to the DNA surface it forms on. So if the relevant DNA segment says "AGCT," the RNA will form "TCGA."[3] Or, to continue our metaphor, hand becomes glove, shoe becomes foot, and vice versa.

This new piece of RNA (called messenger RNA) is the carrier of genetic information. It leaves the nucleus and heads out into the rest of the cell. There it meets with a structure called a ribosome, which "reads" the RNA's code in three-letter chunks, like "AAA" or "GCT" or "CAT." In the ribosome, these three-letter "words" each become a kind of sticky spot for a particular amino acid. Amino acids are the molecules that cells use to build their protein machines.

Another type of RNA (transfer RNA) brings amino acids over to the ribosome, attaching them to the appropriate sticky spots. Each amino acid is then

---

2. Incidentally, the proteins in your Chicken McNuggets are in fact proteins in the broader scientific sense. They derive from a type of long, straight protein that allows chickens to move around before being nuggetized.

3. RNA actually uses a different base than T. This different base is abbreviated as U. We're leaving that out of the main text for clarity, but since you're dorky enough to read the footnote, we thought we'd let you know. U is chemically very similar to T, so you might wonder why RNA bothers using something different. We can't be certain, but there's a good reason why you might prefer U (for "uracil") in RNA and T (for "thymine") in DNA. In short: U is a bit crappier than T. Inside a cell, U takes less energy to make, but is more likely to degrade. This is okay for RNA, which is short-lived. But DNA is the master copy for RNA, so it needs to be more robust.

chemical bonded to the amino acid next to it, forming a long chain. When you assemble these amino acids in a certain order, they fold up into the complex shapes that allow proteins to move around, kick-start chemical reactions, and do all sorts of other tricks needed so that you can continue to do things like eat chips or yell at the news.

Okay, so that's a little complicated. Let's use an analogy.

Think of your DNA as the library of information for how to make machines. In this analogy, if you opened the book of DNA at a random place it might say something like "TNOIJ, EVLAV, POTS, SNEL, EBUT, SNEL, EBUT, TRATS, EVLAV, GNIR, SNEL . . ." and so on for many thousands of pages.

That seems like gibberish. But make a mirrored copy using Silly Putty,[4] and you get "LENS, RING, VALVE, START, TUBE, LENS, TUBE, LENS, STOP, VALVE, JOINT." It's still more or less gibberish, but it appears to have some meaning.

Now, observe that there's a sensible internal snippet: "START, TUBE, LENS, TUBE, LENS, STOP." When you assemble those things in the pre-scribed order, you get a telescope. So to speak, that's the code for a telescope.

But telescopes are boring. Let's suppose you're trying to make something cool. For instance, one of Zach's recent projects was the creation of the world's first single-use monocle. Kelly thinks the project was stupid, but she is wrong.[5]

Somewhere in the DNA library, there's a monocle section that when mirror-copied reads "START, RING, LENS, CHAIN, WRAPPER, STOP." A Silly Putty copy is made and is brought outside the library.

Why would you come outside the library before making the monocle? Lots of reasons, but a big one is that you don't want a machine shop inside your library. If you damage the Silly Putty copy of the "how to make a single-use

---

4. For the young reader: Thousands of years ago, in the twentieth century, there was a toy called Silly Putty—a claylike substance of an inoffensive brown-orange color. Back before fun existed, you would amuse yourself by pressing Silly Putty onto paper with ink on it. When you lifted the putty off the paper, you'd get a mirrored copy of the ink.

5. No, I'm not.

monocle" section, you can still go back to the DNA library and press the Silly Putty in again. And if you want to make lots of single-use monocles at once, you'll want to make a lot of Silly Putty copies and send them off to lots of different factories. If you damage the original book, making new machines becomes impossible. You might make incorrect copies that just don't work. In the worst case, you might make machines that "work," but do something very bad. Something like "START, RING, GASOLINE, LENS, CHAIN, FIRE, STOP."

In any case, once the Silly Putty copy is out of the library, it goes to assembly, aka the ribosome. The assembly people take each word on the Silly Putty and attach glue to it. Then, the parts delivery boy (transfer RNA) brings over the parts and sticks each part on its reference word. Once all the parts are attached properly, the Silly Putty can be discarded or used to make more machines. If nothing goes wrong, the result is a finished single-use monocle, or (if many Silly Putty copies were made) an entire fleet of single-use monocles.

Just so, DNA opens up, RNA copies are made, and those copies leave the nucleus to go to ribosomes where proteins are assembled.

That's a rough-and-ready sense of how DNA does its business, but the actual process can be much more complex, involving things like feedback loops and combining separate chunks of code to make a single machine.

"But wait!" you say. "When people talk about DNA, they always mention *genes.*" Where does that fit in? Well, it turns out a "gene" is conceptually a bit hard for scientists to pin down. Or at least it's hard to define perfectly. This isn't necessarily a big deal. Like, you know how you complain about the economy? Yeah. Now, define it.

One way to think about a gene is that it's a chunk of DNA that appears to do something particular, whether we know what that particular thing is or not. A simple example would be the gene for blood type. Everyone has a section of DNA that, through the above process, determines what blood type they have. Of course, different people have different blood types, but this is just a reflection of the fact that their blood type genes have different code. Type B people and type A people both have a "blood type gene," but the particular codes in that gene vary a bit.

That said, for most features you might identify on an organism, there isn't a single gene. In fact, single-gene (or "monogenic") traits are pretty rare.[6] Even something as simple as hair color or eye color is the product of lots of different genes.

Why should this be? Well, remember, this whole system wasn't designed by anyone. It was the product of billions of years of evolution. Perhaps if humans had built it, each trait would be produced by one particular chunk of DNA, in the same way that each part of a computer is a separate module. But we're stuck with evolutionary history.

6. Some examples of "monogenic" traits are albinism, Huntington's disease, and whether you have the dry or wet kind of earwax. Yeah, there are kinds of earwax. Asians and Native Americans tend to have the dry kind while others tend to have the wet kind.

So, for example, you might be able to find a gene called "GENIUS" that makes its human host 15% more likely to enjoy the classy convenience of a single-use monocle. That doesn't *guarantee* this rare, valuable, and utterly civilized trait will be present in the person—it simply makes it more likely. In conjunction with other genes, GENIUS may increase the likelihood of single-use-monocle-having behavior to near 100%, but if the other genes counteract GENIUS, the unfortunate individual may be bereft of a sense of proper ocular apparatus. Or it may be more complicated—if you have gene A and gene B, but not C, you get one effect, whereas if you have A and C but not B and not D, you get another.

In general, a given gene often only tells a small part of the story.

For our purposes, the important thing to realize is that biology, at the most basic level, is a bit like a messy attic. If you move one thing here, it might cause a collapse over there. This means that if you want to breed for, say, cattle with gigantic horns, the easiest way has long been to just find a bull with big horns and a cow whose dad had big horns, give them a little mood music, and let nature take its course.[7]

This method works well enough, especially if you're not the one with the job of . . . let's call it *concierge*. But for the would-be mad scientist, the complexity of DNA (not to mention its tininess) has imposed limits on how fast and how much we can change biology. Humans haven't been around very long. Very few species have been altered to specifically make things we like. We managed to control the yeast that converts sugar into booze, but why not a yeast that converts sugar into, say, jet fuel? They're both chemicals, right? The DNA just has a chunk that says to make a certain chemical—can't we change that chunk of code to something else?

This is the promise of synthetic biology. If you can create new pieces of DNA and insert them where you like into an organism, you can create biology that *never would have been*. Molecular machines that might turn cancer cells

---

7. The actual process is substantially less romantic, and probably involves a rancher purchasing bull semen and having it shipped in. Incidentally, here are the names of some popular bull semen sellers we found on Google: Bovine Elite, LLC; Universal Semen Sales, Inc.; Sure Shot Cattle Company; Select Sires, Incorporated.

into normal cells. Pest organisms that help kill off their own kind. Or even just general purpose organisms awaiting our instructions. It's life, made to order.

Where Are We Now?

Synthetic biology as we currently know it began in the 1970s. The early methods were complex and cumbersome. Still, a lot of what we take for granted about modern life comes from this era. Human-type insulin (which is probably[8] the kind you want) was only possible to manufacture en masse due to genetically modified *Escherichia coli* (also known as *E. coli*) bacteria and genetically modified yeast. Before that, we used animal insulin, derived from the pancreases of cows and pigs. Animal insulin required the slaughter of lots of animals to get enough insulin, and some people became allergic to the slightly different insulin molecule produced by cows and pigs.

Convincing *E. coli* to make human-style insulin turned out to be a relatively simple process, and it was solved in the late '70s. Other drugs have proven to be trickier.

Fighting Disease

Human beings have managed to wipe out, or at least control, a number of diseases. Others, like malaria, have proven remarkably stubborn. The World Health Organization estimates that in 2015 there were 214 *million* cases of malaria, resulting in 438,000 deaths. This is actually a big improvement over twenty years ago, but there's a long way to go. One particularly good treatment is a chemical called artemisinin.

Artemisinin is a compound extracted from Chinese sweet wormwood plants.

---

8. That "probably" isn't entirely facetious. Some still debate whether we should've moved off animal-derived insulin. It's not available in the United States, but in a number of countries, you can buy an animal-derived insulin called Hypurin. Some patients prefer this version, and it's apparently not terribly expensive compared to insulin derived from modified bacteria.

Artemisinin-based drugs are some of the best treatments we have against malaria, but as every Chinese sweet wormwood enthusiast knows, growing enough of the plant is expensive and time consuming. This is especially bad for malaria because most sufferers live in economically impoverished sub-Saharan Africa.

The availability of Chinese sweet wormwood has varied dramatically over time, creating wild price fluctuations. For instance, the price was about $135 per pound in 2003, $495 in 2005, about $90 in 2007, and about $405 in 2011. When the prices drop, farmers stop growing the plants. This creates the shortage, which increases the price again. And you really don't ever want to have a shortage of antimalaria drugs. Part of why we can't get off this cycle is that the plants take time to grow. Finding a quick way to reliably supply artemisinin should lower the average price, or at least keep the price and drug supply more stable.

Drs. Chris Paddon of Amyris, Inc., and Jay Keasling at University of California, Berkeley, wanted to design a simple organism to create this medicine, so they turned to man's best friend: *Saccharomyces cerevisiae,* aka brewer's yeast. The little fungus that turns sugar into booze.[9]

The challenge is that you can't just make artemisinin. If you think "artemisinin" is hard to pronounce, consider its chemical name: (3R,5aS,6R,8aS,9R,12S,12aR)-Octahydro-3,6,9-trimethyl-3,12-epoxy-12H pyrano[4,3-j]-1,2-benzodioxepin-10(3H)-one.

In order to get some of this stuff, you have to generate a number of different chemicals that react with each other in the right sequence. Over the course of about a decade, their group worked through all the different chemical steps, altering the yeast's DNA so that it would generate the appropriate chemicals to do the appropriate reactions in the appropriate order. Only in the last few years have they finally engineered a modified beer yeast that spits out artemisinic acid, which is easy to convert to artemisinin.

It works well, but the drug is having difficulty competing in the market. The

---

9. Its genus comes from the Greek words meaning "sugar" and "fungus," while its species name comes from the Latin word for "beer." The namers were apparently cool with mixing Latin and Greek, perhaps because of the effects of *S. cerevisiae.*

new technology happened to come on the market just as artemisinin was at a particularly low price, so the old method is currently undercutting the fancy modified yeast. We're not entirely sure who the underdog is here, but it's a good lesson in how technological change is as much about market reality as scientific cleverness.

In any case, malaria is already developing resistance to artemisinin-based drugs in some regions. Thanks a lot, evolution. So what if we could use synthetic biology to stop people from getting malaria in the first place?

The mosquitoes that carry malaria and transmit it to humans often become resistant to pesticides. Mosquitoes reproduce quickly, which means every generation has a lot of chances to produce mutants who can defeat humanity's best weapons. Here's one way we could win the arms race:

Female mosquitoes often only mate once. What if we could trick them into mating with a sterile male? This should mean fewer cute little baby mosquitoes,[10] which means less malaria transmission. An early strategy to make sterile male mosquitoes was to expose them to radiation. This did indeed sterilize the males, but well . . . it turns out that when you expose a guy to a huge dose of radiation, it may increase his chances of sleeping alone.

Later, it became possible to make direct genetic changes to mosquitoes, and scientists made a little addition to the mosquito genetic code. Mosquitoes with this added gene require the antibiotic called tetracycline, or they die. You give them tetracycline when they're born, and then they go off and mate. They then produce children who will die without tetracycline. Those children don't have a scientist standing by to deliver the antibiotic christening, so they die off before they can make another mosquito generation.

---

10. Dr. Scott Solomon of Rice University suggested we point out that baby mosquitoes aren't miniversions of the adults, but rather "wormlike larvae that writhe around in standing water." So not so cute, even for mosquitoes.

Okay, so this works in the short term, but it's very expensive. When you introduce a gene that results in all of its descendants dying, that gene doesn't last too long in a population. The mosquitoes bounce back in a few generations. This means that if you want to keep mosquito populations down, you have to keep introducing carefully gene-manipulated mosquitoes over and over.

Unless you have a gene drive. Here's how it works:

When a mommy and daddy love each other very much, they turn out the lights and combine their DNA. If everything goes terribly wrong, they have a baby. Now, suppose there's a single gene for color of nose hair.[11] Mommy gives you a gene for black nose hair while daddy gives you a gene for vibrant orange nose hair. Later, you will produce offspring whose nose hair color is influenced by the genes you got from Mom and Dad.

But now imagine Dad's orange-nose-hair gene was not an ordinary gene. In addition to coding for orange nose hair, it also *destroys* the other partner's nose-hair gene. When this happens, the DNA fixes the destroyed gene by copying

---

11. There isn't. Or should we say *there snot.**
*Sorry.

the orange-nose-hair gene. So now you have two of Daddy's orange-nose-hair genes and none of Mommy's black-nose-hair genes.

What happens? Well, you definitely end up with orange nose hair. But then . . . something more sinister and orange takes place. All of your siblings also have orange nose hair. And so do all of your children. And their children! The gene "drives" its way through the entire population. This is true *even if* vibrant orange nose hair (inexplicably) makes you less sexy. You might have fewer kids, you ginger-nosed freak, but *all of them* will have the selfish gene.

So gene drives mean that you can impose synthetic biology on an entire wild population. Dr. George Church of Harvard and others were able to put multiple gene drives for malaria resistance into mosquitoes. That way, instead of having to release modified mosquitoes every season, you could potentially have a one-time release of mosquitoes with multiple malaria-resistance gene drives. Even if malaria resistance makes them less attractive mates, the gene should spread through the population, growing exponentially as it goes.

According to Dr. Church, the goal is complete eradication of malaria. But releasing an organism with engineered DNA into the wild in order to kill off an entire species (even if the species was a parasite) was bound to raise some scientific hackles. A panel at the U.S. National Academy of Sciences thinks the technique is promising enough that they gave the go-ahead, but they want to see a lot more research in this area before thinking about releasing these mosquitoes into the wild.

Medical Diagnostics and Treatment

When your dog runs into your house, it can't tell you that it's been rolling around in mud and eating squirrels. Somehow, you just know. The body of the dog holds a record of the day's journey in its smell and appearance. Some scientists wondered if the same could be true for bacteria. This would be useful, because you could send the bacteria on a magical journey through your digestive tract. They could put together a little scrapbook of the trip, then give it to your medical provider when they . . . emerge. It's not the most delightful image, but it might be preferable to the current method if sticking a camera in there.

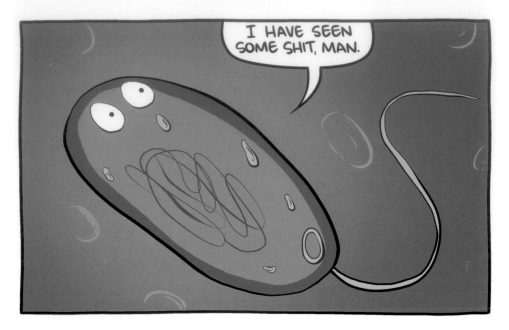

Dr. Pamela Silver and her lab at Harvard Medical School had a thought: Could you create a synthetic mechanism that could capture information into a bacterium's DNA and recover that information later? Basically, can you get bacteria to "observe" their environment and then change in some way that tells you what they saw? The answer is yes. *Duh.*

Here's the idea: Two cells of the same type, with the same starting DNA, can have variations that they acquire from their environment. For example, there may be molecules attached to their DNA that change how it codes for things, or there may be some kind of chemical feedback loop that results in higher or lower expression of some particular gene. And, in some cases, a cell may pass its acquired alterations to its offspring.

These changes, if they persist in a form we can decipher, could tell us about what the bacterial cell "saw" on its trip. The problem is that bacteria aren't naturally designed for this purpose. It's sort of like the dog from earlier—if it ran to the neighbor's house, suddenly developed intelligence, killed the neighbor for his money, took up online poker, lost it all on a bluff gone wrong, went mad and lost its intelligence, then ran through some mud and came home, well . . . all you'll see is that there's mud on the carpet.

With a dog, you can just strap a camera on. With bacteria, you can't. It's too small, and the sun doesn't shine where it's going anyway.

In DNA, you can find a sort of chemical loop[12] in which the DNA creates a molecule, and the molecule tells the DNA to do it again. It'd be sort of like if you made a sign that said: "WHEN YOU READ THIS SIGN, MAKE A COPY OF IT, THEN READ IT." Once the loop triggers, you keep making signs forever.

In principle, such a chemical loop should be able to function as memory. To continue the sign analogy, suppose you had a mental program that went something like this: "When your pants fall down, make a sign that says 'WHEN YOU READ THIS SIGN, MAKE A COPY OF IT, THEN READ IT.'" If we later saw you making endless signs, we could conclude that your pants had likely fallen down without ever having to make the necessary observations.

DNA works in a similar way: Once the loop is turned on, it'll just keep going. And the active loop will pass down through generations of cell lines, at least for long enough that you can recover useful information from it.

Dr. Silver's lab has already done this. They've created synthetic DNA loops that are inserted into bacterial DNA. When the bacteria experience certain conditions, their loops activate. Because you are creating the loops synthetically, you get to decide what chemicals they're creating. This means you can decide on a chemical that's easily detectable by, for example, shining when exposed to certain types of light.

To give you a sense of why this might be useful: tumor cells often experience repeated oxygen deprivation because they grow so quickly that they don't get a sufficient blood supply. It turns out that repeated oxygen deprivation produces a detectable chemical signal in tumor cells. So Dr. Silver's idea is that you can put programmable memory cells into someone's body, then check them later to see if they've detected areas of low oxygen. If they did, it's possible they found solid tumors.

---

12. If you must know, it's called positive transcriptional autoregulation.

This method is still in an early phase, but the clinical applications could be incredible. Once you have this general method for creating cell-sized programmable sensors, the door is open for researchers to program the cells to detect all sorts of things.

When you pair this with other work Dr. Silver has done, things get really interesting. For instance, Dr. Silver and her colleagues published a paper in 2016 with the title "A Tunable Protein Piston That Breaks Membranes to Release Encapsulated Cargo." That is, you can program bacteria not just to listen for problems, but to deliver treatment.

This method could have applications from cancer drug delivery to treating irritable bowel syndrome. It is also targeted. Right now when you have inflammation of the gut, you take aspirin orally, releasing the chemical into your entire body, most of which is not inflamed. In principle, you could create a bacteria species that holds on to aspirin until it detects the chemical signature for inflammation. If this method could be generalized, as Dr. Silver's work suggests, perhaps all sorts of medicines could be delivered right to the relevant target, maximizing effectiveness and minimizing side effects.

Organs from Nonhumans

Scaling up a bit, it may be possible to synthetically modify large animals to make things we like. In our chapter on bioprinting (stay tuned) we talk about the challenges of quickly building whole organs from scratch. But animal bodies already know how to create organs. So next time you see a pig, maybe you should visualize a 3D printer for kidneys instead of a plate of bacon.

It turns out that pigs are particularly good for this sort of thing because their organs are similar in size to ours. Transferring organs between species—which has the ominous name xenotransplantation—doesn't have an excellent history of, you know, working.[13] Most of the research has focused on transferring between nonhuman animals, like from pigs to baboons. We still haven't figured

---

13. Well, pig heart valves have been used to replace human heart valves with great success for over a decade. But scaling up to replacing something like an entire heart turns out to be much more difficult.

out how to keep the immune system of the recipient animal from killing the donor organ, but some progress is being made. One thing that would make this all much easier would be if the organs from the donor animal "looked" to the immune system more like organs from the recipient animal.

So just like we modified yeast to create malaria drugs, we should be able to "humanize" pig organs. Part of how your body knows you just got a pig heart instead of a human heart is that the pig heart contains and creates a lot of molecules that are similar to the human version, but different enough to be recognized as foreign. For example, there's that pig version of insulin we mentioned earlier. By altering pig genetics, we should be able to make pigs whose organs are molecularly very similar to human organs.

Scientists recently announced that a pig heart had been kept alive in a baboon for over two years,[14] using this method along with drugs to reduce the immune response. Some readers may feel squeamish about putting pig organs in humans, but we suspect they wouldn't feel that way if they urgently needed a liver.

A more serious concern is that we'll accidentally pick up brand new diseases from pigs. Dr. Luhan Yang and her lab at eGenesis are working to fix this. The problem is that pigs contain "porcine endogenous retroviruses," which are (yes, really) referred to as PERVs. PERVs are found in pig DNA, and they release particles that infect humans. Humans don't want PERVs inside them,[15] so they employed a new technique called CRISPR-Cas9 (discussed shortly)—to cut out the pig PERVs. This doesn't eliminate all risk of a disease-jumping species, but it eliminates one of the scarier possibilities.

---

14. The pig's heart was sewn into the baboon's abdomen, but did not replace the baboon's heart. Still, a step in the right direction.
15. Sorry.

Fuels

Cells are probably the best chemists there are.

· Dr. Pamela Silver

In 2009, Dr. Dan Nocera of Harvard discovered a relatively cheap catalyst that could be placed in water. When it got a little energy, it would split water (H_2O) into just the H and the O.[16] This catalyst was potentially a big deal because when you split water into hydrogen and oxygen, you have an excellent form of stored energy. When you bring these elements together and apply some energy, you get a big explosion when they combine back into water. Hydrogen fuel cells essentially do this trick but without the explosion part.

So the idea is that you can make cheap, clean fuel cells by using heat to split water and then bringing the split parts back together when you need energy. This is really just a simplified and controlled version of how plants make energy. According to Dr. Silver, "The process of photosynthesis, which is one of the most amazing things nature does, is to harvest sunlight and use that sunlight for energy to make stuff. That's the basis of life on earth. . . . One of the key reactions in photosynthesis is called the water-splitting reaction."

But for Dr. Nocera, things didn't work out as planned. The device worked well enough, but hydrogen fuel cells never caught on as a way to store energy. Making things worse for him, but great for everyone else, regular old solar power cells got a whole lot cheaper, making his product less exciting. Dr. Nocera's idea was shelved for a time. But harnessing water splitting has a lot of potential. You've got this ultracheap way to split water, but the way you get energy out of it is cumbersome. Then, Dr. Silver had an idea.

Her lab introduced a genetically modified bacteria that could take the hydrogen and oxygen, combine it with carbon dioxide, and convert it into isopro-

---

16. It's actually the stable H_2 and O_2 forms, but you get the point.

panol. Isopropanol is a fuel that can be separated from water. The result is a system in which the inputs are a metal catalyst, water, bacteria, and carbon dioxide, and the output is a chemical you can use to run your stove.

This is kind of wild if you think about it. You put the right material and biology in water, you let it get some light and heat, and then you start seeing fuel forming in the container. And because you're mimicking life and extracting CO_2 from air, your fuel is much more environmentally friendly.

Dr. Silver told us, "Our contribution was . . . to make that process as efficient as probably the best photosynthesizer, which is algae. Actually, we beat algae now. In the original paper I think we said we were beating plants; we've now beat algae."

They are currently seeing if the process can be made cheap and scaled up.

Another group is working on a way to go from switchgrass to jet fuel. If you've never heard of it, switchgrass is a tall green plant that's all over North America. It's a hardy grass that grows quickly and densely, even in bad soil.

In case you don't remember high school botany, cellulose is one of the main structural components in plants. Cellulose is a very long chain of sugar molecules. So perhaps you're now thinking, "Why don't trees taste good when I lick them? I keep licking all sorts of plants but they're almost never sweet!"

First, stop it. Second, cellulose chains are pretty hard to break down into tasty sugar molecules. Unless you have specially developed enzymes to break down cellulose sugar, you can't digest it. This is why cows have complex digestive systems—they're doing God's work of converting stubbornly hard-to-digest grass into beef.

But doing your business inside a cow is never a good way to go. So, Dr. Aindrila Mukhopadhyay's group at the Joint BioEnergy Institute created bacteria that can convert renewable plant resources (like switchgrass) into d-limonene, a precursor to jet fuel. Her group's modified bacteria can take pretreated switchgrass, break the cellulose into little sugars, then turn those sugars into d-limonene.

They are hoping to get the process to the point where the bacteria spit out straight-up jet fuel, but getting to d-limonene in one pot already cuts out a lot of steps you normally need to make bio-jet fuel. And, because the carbon source is the airborne CO_2 that the switchgrass converted into cellulose, in principle, you could get your jet fuel without adding much CO_2 to the atmosphere.

Biofuels like this have enormous potential to reduce our reliance on petroleum-based products, but so far cost has been a serious issue. Petroleum prices continue to be quite low, so it may be some time before we're using weeds to fly to Amsterdam.

Environmental Monitoring

Dr. Silver was able to give cells the ability to remember and report back on their experience inside the body. Could we do the same in open environments?

Dr. Joff Silberg and Dr. Carrie Masiello are a husband-and-wife team of professors at Rice University. He is a synthetic biologist. She is a geologist. But somehow they managed to move past that and find love.

Dr. Masiello studies biochar, which is made when plant matter gets baked at a high temperature in the absence of oxygen. The creation of biochar sequesters carbon that would otherwise end up back in the atmosphere, and it is frequently added to soil to increase plant growth. We don't know precisely why it helps plant growth, but it may be that it alters the composition of microbes in the soil. Dr. Masiello wanted bacteria that could report back to her on what conditions were like for microbes living in soil with and without biochar. She asked Dr. Silberg to make her a synthetic microbe for Valentine's Day. Yes, really.

Dr. Silberg created bacteria that release gases that aren't commonly found in soil. So by putting the synthetic microbes in soil, then monitoring the gas release, we can "eavesdrop" on microbe behavior instead of grinding them up for analysis.

CAN YOU HELP ME EXAMINE SEQUESTERED CARBON?

because you've already sequestered my heart

(seriously though)

Most of us aren't quite so romantic about gassy soil microbes, but this technique could be extended to environmental contamination. Groups have already modified bacteria to glow in the presence of arsenic and water. The more arsenic, the more glowing. It's like a nightlight that runs on poison.

Potentially, a technique like this could be used to find and monitor toxic environments. Thanks to advanced synthetic biology, you can already write bacterial programs that are more complex than just "Glow if you detect toxins."

A major hurdle for this field is that supertoxic environments are bad for bacteria too. So when you don't see glowing, you think it's because no arsenic was detected, but *actually what happened is all of your bacteria are dead.* Scientists are working to solve this problem. One idea is to use particularly hardy organisms that are already used to toxic environments.[17] Another idea is to create more complex signaling mechanisms, along the lines of "Red light means bad, green light means good, no light at all means MOTHER OF GOD RUN FOR YOUR LIFE."

If people decide they're okay with synthetic bacteria running around in the world, you could do continuous monitoring of conditions just about anywhere. If the soil in a particular area started glowing green, you would know there was arsenic. Blue could mean toxic levels of mercury. Yellow could mean lead. Basically, if you come upon a mysterious cove where nature enrobes herself in prismatic illumination, don't drink the water.

Generalizing Synthetic Biology

All these things are neat, but they're also really hard. There have been ways to modify genetics directly since the 1970s, but the methods are difficult, expensive, and time consuming. Or, at least, that *was* the case. In just the last few years, a new method has come on the scene that promises to change everything.

A group led by Dr. Jennifer Doudna from the University of California, Berkeley (and the Howard Hughes Medical Institute), and Dr. Emmanuelle Charpentier of the Max Planck Institute for Infection Biology discovered a way to make molecular scissors, thanks to a quirk in the way bacterial immune systems work. The system in bacteria is called CRISPR-Cas9. If you didn't guess, that first acronym is short for "clustered regularly interspaced short palindromic repeats." Just pronounce it "crisper," like the drawer in your fridge.

A naturally occurring bacterium doesn't have memory storage in the same sense that you do. It can't see, hear, or think. But bacteria are able to fight off viruses they've encountered before. Somehow, they "remember" viruses and attack them.

---

17. You hear that? That's the sound of us resisting the urge to make a New Jersey joke here.

Here's how it works: When a virus infects a bacterium, it injects bits of genetic material through the bacterium's cell wall. These bits of genetic material try to take over the machinery of the cell in order to make more virus particles. But the bacterium has a protein called Cas, which can fight off the virus. When Cas is successful, it takes part of the defeated virus's genetic material and adds it to a special section of the bacterial cell's DNA. This gives the bacterium a way to remember the virus.

Later, when that bacterium bumps into the same virus, it "recognizes" it using the stored code and then cuts the virus protein in the recognized place. The bacteria isn't cutting in the recognized place out of a sense of poetic justice—it's just that when someone attacks, cutting them in pieces is a pretty good defense. But the result is a handy tool for humans—Cas always snips at a certain genetic location. *Targeted molecular scissors.*

And here's the cute part: In a healthy cell, when its DNA gets snipped, it attempts to repair itself by joining the two ends back together. Before the repair happens, you can slip in new molecules that fit in the gap. The DNA heals itself up, and BAM! You've just selectively introduced new code into a cell's DNA at a spot of your choosing. And you did it into a living cell.

The lab groups of Dr. Feng Zhang at MIT and Dr. George Church worked out methods to use CRISPR-Cas9 on mice and humans. So, as of 2013 or so, you can run around in cells from all sorts of organisms, snipping DNA out and sticking DNA in willy-nilly. What shall we do? Play God? Besiege the ancient vale of Nature with the iron cannons of Science?! Don't mind if we do!

Of course, as the Garden of Eden story tells us, when you play God with some preexisting organisms, they don't always behave right. Until we can make whole organisms from scratch, we've got to twiddle with DNA inside creatures that nature made. But we don't have to play nice with nature.

The Simplest Organism

Dr. J. Craig Venter famously raced Dr. Francis Collins's National Institutes of Health to decipher the human genome. Now he's on to bigger things. Just to give you a flavor of Dr. Venter, he once responded to the writing prompt "What *Should* We Be Worried About" with an article that began, "As a scientist, an optimist, an atheist and an alpha male I don't worry."

See, this is exactly the kind of person you want if you need someone to play God. He works at a place that happens to be named the J. Craig Venter Institute. Aside from being fearlessly alpha all day every day *to the max*, Dr. Venter's team is working on creating the simplest organism possible.

Their idea is that if you have an extremely simple organism, it should be relatively easy to tell what'll happen when you alter its DNA. You'll have a sort of blank canvas for new genes, so scientists can figure out the effect of their changes much more quickly.

They began with an organism called *Mycoplasma genitalium*, so named because it is found in human genital and urinary tracts. In addition to being conveniently located, this organism has an extremely short genome. They started snipping out and discarding more and more genes to see what was essential to survival. Sometimes losing a gene killed the organism, but sometimes it didn't. After a lot of work, and after (sadly) switching to a related species with a less funny name,[18] they arrived at an organism with only 473 genes.[19] Humans, by comparison, have about 20,000 genes. Venter's group dubbed the new organism *Mycoplasma laboratorium*.

Dr. Venter is an atheist, but just in case there is a God to piss off, he named the latest version of this organism Syn 3.0. Investors are already signing up to see what Syn can do for them.

---

18. *Mycoplasma mycoides*. It has a slightly larger genome, but reproduces much faster than *M. genitalium*, which makes it easier to work with in the lab.
19. Fun fact: Over 30% of the genes that they kept have no known function, but were apparently critical to the organism staying alive. So even with alpha-wolf Venter in charge, we're tinkering with a language we don't yet fully understand.

Biohackers and Standard Parts

For those of us who aren't quite as excited by a secretive genius creating a proprietary form of life, there is also a grassroots approach to synthetic biology.

A competition called iGEM (International Genetically Engineered Machine) happens annually, and pits students (including high school students!) against one another to see who can create the most exciting genetically engineered organism. In 2015, teams created (among other projects) a biosensor that worked with your iPhone and could detect heavy-metal contamination and date-rape drugs, an organism that secretes chemicals to adjust the freezing point of the water it's in, a cheap and fast test to determine if a cancer has metastasized, and a biosensor that quantifies the purity of heroin.

The teams also put any "parts" that they create in a "Registry of Standard Biology Parts," which is freely available and can be added to by people who did not participate in iGEM as well. In other words—open-source biology Legos. You can order these parts and do synthetic biology research if you have the equipment, which is becoming increasingly available at "biohacker" spaces. So your neighbor could be the next person to solve our energy crisis, or cure a disease, or write "KICK ME" in your skin with bioluminescent chemicals. If you lived near MIT this was probably already true, but soon it may be an experience for everyone to enjoy.

Concerns

Tinkering with the language of life. What could go wrong? An early mantra of the Internet was "information wants to be free." That sounds nice, but it's a problem if the information is how to make smallpox from scratch.

Ultimately, synthetic biology should give humans the power to have organisms made to order. As that technology becomes cheap, the ability to bring back diseases for which we no longer vaccinate might become something you could do on your desktop.

Consider smallpox: After 1980, we stopped vaccinating because it was mostly gone.[20] This disease may have killed half a billion people in the twentieth century, and most living people have no immunity to it. If synthetic biology became easy, what would stop a rogue biologist (or just an angry geek) from bringing it back?

Scarier still is the possibility that a disease like smallpox could be modified to make it spread more rapidly and be more lethal. A biohacker could in principle design the disease to elude all known therapies. We also know that some diseases may affect human behavior. For instance, flu vaccines apparently make human beings more social, perhaps to the benefit of the disease. A disease designer could potentially make society-level behavioral alterations via a subtle pathogen.

At the moment, the companies that make made-to-order DNA keep an eye on the orders submitted by their customers. But as DNA synthesizers get cheaper and cheaper, could this become the kind of thing you could do at home? The best hope may be that the power to create diseases will come along with the power to fight them. But prevention by some means is probably better than an arms race in which human beings are the battlefield. One other bit of cold comfort is that bioterrorism is rare, probably because it's tough to control. Terrorism, by definition, serves the actor's political goals, but very few people's interests are served by creating a life form that might easily infect and kill their own people.

---

20. Samples still reside under lock and key at disease research centers in Russia and the United States. You know, except when it's not under lock and key . . . like when someone at the Centers for Disease Control in the United States forgot some smallpox vials in a closet. The vials were rediscovered many years later. Whoops.

Ecologists worry that synthetic organisms might become invasive by accident. This could be scary if we're using them to mass-produce industrial chemicals. Having a pot of bacteria that churns out jet fuel is fine, but what happens if the bacteria gets loose and starts doing its thing in a river? The hope is that these bacteria, which are designed to work under particular and unusual conditions, wouldn't do well in the wild. But bacteria can exchange genes with each other, and they evolve rapidly. Scientists are working on ways to prevent synthetic bacteria from exchanging genes, but there's no way to completely ensure safety.

And what about the organisms we're creating that we specifically *plan* on releasing into the wild—the gene-drive mosquitoes from earlier? Dr. Silberg has some reservations about the pace at which we release synthetic organisms into the environment. "I want to see people doing molecular gene stuff interacting with ecologists a lot more to make sure that we fully understand the potential environmental impacts as we consider the potential benefits. Because there's a whole field of ecology where people think about what wacky things people have done in Australia over the years, the catastrophes that have come from not really thinking about what this means from an ecosystem perspective." For example, we introduced cane toads in Australia to control a native beetle that was a pest on sugarcane plants. The introduced toads reproduced like crazy and started spreading across the continent. These toads produce a toxin that kills their predators, and so as the toads spread they killed native predators (and some unfortunate pets).

This problem gets a little more personal when we consider introducing synthetic organisms into our own bodies. Even if they're carefully engineered, there might be some risk of a mutation that makes them dangerous. But, as Dr. Silver notes, dangerous mutations are already a possibility with the nonsynthetic bacteria inhabiting your body.

In the next chapter we discuss using CRISPR-Cas9 to fix genetic disorders in human beings. Most of us are cool with the idea of using the techniques of synthetic biology to cure diseases in adults, but some scientists are also proposing using CRISPR-Cas9 to cure diseases in human embryos, making changes that would be passed on to subsequent generations.

There are some who argue that the benefits outweigh the risks. But where would we stop? If we're able to modify human embryos, what's to keep us from making designer babies? Adjusting hair, eyes, skin, perhaps even IQ, might be an option in another generation or two. And it won't be an option for everyone. In a world of haves and have-nots, one group may be able to produce uniformly superhuman children while another group has to deal with genetic disorders that are easily fixable.

At the time of this writing, scientists in the UK are allowed to modify human embryos, while U.S. scientists are not. In China, CRISPR-Cas9 was used to modify human embryos, and the results were pretty abysmal. Lots of things went wrong, including unexpected mutations popping up. Remember, we don't yet know what we're doing here. Even if a designer human were successfully created, we don't know how its genes would affect future generations.

How It Would Change the World

It's the most amazing time. You no longer have the burden of really tedious experiments; you're just limited by imagination.

• Dr. George Church

Synthetic biologists want to make their knowledge open and easily accessible. In less than a single lifetime, we've gone from wondering about the structure of DNA to reprogramming it with its own machinery. In the more distant future, this could have some wild ramifications. Like, storing memory in DNA.

How? Well, remember that all the memory in your computer is just an arrangement of 0s and 1s. DNA is just an arrangement of letters—A, C, T, G. In a certain sense, it's like an ultracompact version of the strips of memory tape computers once used. By synthesizing DNA with the right code, you can store up to 10 billion gigabytes of data in a space smaller than a drop of water. That's fifty million copies of *The Lord of the Rings* movies, or half a copy of Windows 10. And DNA is an extremely stable molecule, with a half-life of about five hundred years, meaning in about five hundred years half of the information will be degraded.

We don't do this outside of the lab yet, because writing out custom strings of DNA is expensive. Right now, it's about 10 cents or less per letter. For reference, the human genome has about three billion letters. Scientists hope that demand for DNA synthesis will drive down the price. Dr. Church proposed a follow-up to the Human Genome Project, in which we synthesize the complete human genome as a way to start driving these costs down.

In the future, we might take synthetic organisms to space. As you're now well aware, getting stuff to space is expensive. If you had bacteria to manufacture products and recycle waste, you could make much better use of available resources. This becomes especially important if you want settlements on other planets and moons. Synthetic bacteria could be specially designed to work with any local environment to make whatever resources you need. And because those bacteria would be able to replicate, you'd only need to take a few of them with you.

So-called GMOs (genetically modified organisms) have gotten a somewhat bad reputation as "Frankenfoods" in the popular discourse. Our view is that if a Frankenfood can provide more calories and vitamins to poor communities, we should call it a win. But even if you want your tomatoes to be untouched by the hands of bioengineers, you might appreciate that GMOs can also give us greater access to medicines and clean fuel.

As we learn more, the line between synthetic biology and nanotechnology becomes meaningless. We are pursuing smaller and smaller machines, and biology just happens to have had an extra four billion years to learn how to manufacture them.

Recently, scientists created a new form of DNA. Regular DNA, as you now know, has four chemical letters—A, C, T, G. The new DNA has two new letters—X and Y. This is completely alien—it is something that would be mind-blowing if we'd found it on another planet. But instead, we've created it in the lab. And it's not just neat—it comes with a lot of possibilities.

Normal DNA can make 20 amino acids—the building blocks of life's molecular machines. This new DNA can make 172. The number of possible proteins it could make includes many that have never been made in nature.

This is the promise of synthetic biology—not just changes to life as we know it, but the creation of life as we might imagine it.

Nota Bene on De-extinction

Somewhere between two hundred and two thousand species are currently going extinct each year, and a lot of that is our fault. As an analogy, you can imagine humanity as a gigantic pair of jaws gobbling up everything beautiful and excreting the tawdry spectacle we call civilization. In fairness, civilization created nachos, so you know . . . trade-offs.

We *could* stop destroying habitats and introducing invasive species, of course, but so far our track record hasn't been great.

Because ecosystems are complex, destroying one species—even one that appears to have a small effect on its community—can result in the extinction of others. This collateral extinction can cause further extinctions, and so on. But what if we could stop the ripple of destruction by bringing back recently lost species? Preserving these diverse environments intact is one of the goals of scientists who study how to resurrect extinct species. Yep. It's about ecosystems. Certainly not about riding dinosaurs and eating mammoth-burgers.

Okay, so you want to bring back long-dead animals. It's not as easy as it looked in *Jurassic Park*. In order to have a chance, you have to have the genome of the lost organism. Humans have only known that DNA carries heredity for a few generations, so it's not really fair to expect cavemen to have preserved mammoth DNA in nice climate-controlled containers for us. But here and there, by accident, nature has preserved ancient DNA.

The longer DNA is exposed to the elements, the more it degrades. Right now, scientists think that after about a million years of degradation, DNA can't be recovered. This fact precludes almost all of the most awesome animals. But it does leave a lot of possibilities: mastodons, dodos, saber-toothed cats, and maybe even near-human relatives, like full-blooded Neanderthals.

We talked to Dr. Beth Shapiro at the University of California, Santa Cruz, to find out how we could get a pet mammoth. You know, for ecosystems or whatever.

First, you locate as much of the mammoth genome as you can. This is not a small task. Even if a mammoth was well preserved in the Russian tundra and you get a good DNA sample, only a small percent of it likely belongs to the mammoth. The rest of the DNA belongs to the microbes that lived in the soil, the scientist who didn't take enough care when moving the carcass, or the cavepeople who thought to themselves, "Screw you, future geneticists," and spat on the dead mammoth during the Ice Age.

So you have to filter through your data to find the small quantity of DNA that actually belonged to the mammoth, and then do your best to figure out where in the mammoth genome those DNA sequences actually belong. With luck and patience, you can assemble a mammoth genome that is very close to what you would've found 20,000 years ago. But, you're not going to be able to get the entire mammoth genome. Inevitably,[21] some pieces will be missing, and the pieces you *do* have will be so small and scattered that you're going to need a cheat sheet to figure out how to put them back together.

Now you take the genome of the Asian elephant and compare it with what we have of the mammoth genome. Asian elephants are closely related to mammoths, so you can take a full Asian elephant genome and splice in mammoth genes wherever appropriate. These genomes probably differ by about 1%, but that's still a lot of genome tinkering. Like, a lot more tinkering than we can easily do at the moment with our current techniques. Near-term attempts at mammoth de-extinctions may be more elephanty[22] than desired.

But now say you've got the mammoth DNA. Well, DNA doesn't just spontaneously turn into a full-grown animal. If it did, there'd be a lot more teen fathers. You insert the mammoth DNA into an elephant egg and impregnate an Asian elephant mom with a woolly mammoth baby. This further compounds the elephantishness problem, because a modern elephant may have crucially different womb conditions, due to genetics, diet, hormones, and so on.

And, once born, the mammoth needs a modern elephant's microbes. "So," says Dr. Shapiro, "it's born, it eats a little bit of elephant poo, because elephants do that to establish the community of microorganisms that live in the gut that can then be used to break down the food that elephants and mammoths eat. It

---

21. Not all scientists think it's inevitable that some pieces of the woolly mammoth genome will be missing. For example, Dr. George Church is working on woolly mammoth de-extinction and thinks we may one day have a complete woolly mammoth genome.
22. Measured in units of elephantishness per liter.

will then have an elephant's microbiome." Welcome back to Earth, long-dead creature! Welcome back, sole representative of your kind! Now, eat some poop.

Voila, you've got a mammoth, albeit it with some elephanty features. If the goal was to 100% recreate the lost mammoths, you probably fell short. But if the goal was to create an animal that played an important role in an ecosystem before it was lost, well then, maybe an elephantlike mammoth is pretty good.

As Dr. Shapiro tells us, "An ecological replacement really can replace these missing ecological interactions that have disappeared because of extinction, in a way that revitalizes that ecosystem and saves living species from going extinct. This is what I see as the power of this approach. We don't have cold-adapted elephants, but we could use synthetic biology to make them. In that way we could replace this missing component of this ecosystem and reestablish these rich grasslands that used to live in Siberia, creating habitats for things like saiga antelopes, and wild horses and bison."

Let's assume for a second that our technology will one day advance to the point where cloning a mammoth is possible. Where is this mammoth going to live? Mammoths used to live in Siberia, but will the Siberians welcome mammoths back with open arms? Perhaps not. Gray wolves were reintroduced to Yellowstone National Park in 1995, and the ranchers in that area haven't exactly been thrilled. There have been lawsuits, and ranchers have gotten in trouble for shooting wolves to protect their cattle.

But it turns out there is already some habitat in Siberia just waiting for mammoths to return. Dr. Sergey Zimov has been waiting for them, and has put aside land on which he hopes mammoths will one day roam again. He also probably has tourism on his mind, as he has named the land Pleistocene Park.

Maybe at this point, you're thinking, "So there's *no* way to bring back dinosaurs?" We had an expert on the phone, so we asked her.

"Using a computer, we could reconstruct what the genome sequence of the ancestor of all living birds, avian dinosaurs, looked like, and that would in essence be a dinosaur. It wouldn't be a T-Rex or a brachiosaur or a velociraptor because we don't have any DNA from them, but it would have been something that lived contemporaneously with dinosaurs and was the ancestor of all living birds. We could use synthetic biology to gradually swap out modern bird (living dinosaur) genome pieces with this ancestral computationally inferred bird (dinosaur)."

Well, we didn't spend our childhoods dreaming of riding a computationally inferred bird ancestor, but . . . close enough.

SECTION 3

You, Soonish

9.

Precision Medicine

*Everything That's Wrong
with You in Particular—
a Statistical Approach*

Before the nineteenth century, getting medical treatment kinda sucked. Anesthesia consisted of a shot of whiskey (or two), and medicines consisted of traditional cures, like bloodletting and hedgehog grease, with no empirical validation. In the modern world, we rely on the fact that practicing doctors have a deep alliance with scientific researchers. Researchers test treatments to see what works, and doctors weigh the evidence when deciding what to do for patients.

A good doctor is a sort of a Sherlock Holmes of your body. Whether you are sick or well, your body is constantly giving off little clues to your internal state. Some of these clues are quite obvious—if there's a giant hole in your head, the doctor can be pretty sure of what ails you.

Other cases may be very difficult to pin down. For example, most patients diagnosed with mononucleosis only get that diagnosis after many other incorrect ones.[1] This isn't really the doctor's fault—mononucleosis is a viral disease that is quite long lasting, and its symptoms are general things like fatigue, headache, and sore throat. It's sort of like a constant mild hangover, which makes it very hard to differentiate from baseline human existence. But if doctors first checked what molecules were prevalent in your bloodstream, they might diagnose you with mono instead of a cold.

What doctors have long called "symptoms" or "signs" more computationally inclined researchers now call "biomarkers." A biomarker, broadly defined, is just about anything that tells us about someone's internal state, usually in reference to whether something in there is going wrong. Most commonly, biomarkers are

1. And other patients get diagnosed with mononucleosis when they really have something else. We were made aware of one case where a person was repeatedly diagnosed with mononucleosis over a period of four months before finding out he had secondary syphilis. Whoops.

traditional symptoms as well as chemical cues in your body, but some research-ers think the term can be broadened to encompass behavior patterns, like what Web sites you browse or what images you post online. So to speak, if you have a computer model called "How Am I Doing?" a biomarker is anything you might input into that model that would help it find an answer.

Just as the coming together of science and medical practice brought about modern medicine, the coming together of medical science with molecular anal ysis, data science, and machine learning may bring about a new paradigm, which is coming to be called precision medicine. In the future, you may get medical diagnoses that are determined quickly and correctly from thousands of biomarkers, followed by treatments that are tailored to you in particular. This means you will live longer, live healthier, and—if the detection systems get cheap and easy enough—you don't spend nearly as much time wondering if that bump on your right butt cheek is cancer. Not to mention, if diagnosis and treat-ment of disease becomes a matter of computer power, it has potential to (for once in our lives) drive down the cost of health care.

In a single drop of your blood there is a staggering amount of information. There may be chemical biomarkers associated with looming heart failure. There may be genetic code from an undetected solid tumor. There may be hormonal biomarkers that tell us that you're more stressed than you realize.

A deep knowledge of these subtle biomarkers not only leads to better diag nosis, but it also suggests novel treatments. If you have cancer, that cancer has certain genetic mutations we can locate. Having located those mutations, we can pick the best treatments, or go "off the shelf" to pick a treatment that might work even though it's not designed for the disease you're experiencing. With the very latest methods, we might even be able to create a general technique for dealing with any disease arising from genetic mutation.

As we get more data on all the variety and complexity of human bodies, and as we get better at analyzing that data, we approach a time when a computer delivers a perfect diagnosis and selects the ideal treatment method. This dream may seem far off—and at least some aspects of it are quite a ways away—but remember, the human body is of finite complexity.[2] Every advance brings us closer to the finish line.

Where Are We Now?

To get a sense of how the field has changed in the past fifty years, Kelly sat down with MD Anderson Cancer Center's Dr. John Mendelsohn.

Let's imagine, hypothetically, that you didn't do your research on Dr. Mendelsohn because a trusted colleague told you this doctor was the perfect person to talk to about precision medicine. If you were that foolish, you would still quickly learn how important Dr. Mendelsohn is because you would look up his address *after* arriving at the MD Anderson Cancer Center campus and learn that his office is in the *John Mendelsohn Faculty Center Building*. At that point you would, hypothetically, do a panicked Google search for "John Mendelsohn," and discover that he is, in fact, the former *president* of MD Anderson Cancer Center.

You would take a hypothetical deep breath, discover that your armpits were now soaked with sweat, and resolutely knock on his door.

Fortunately for hypothetical you, Dr. Mendelsohn is a friendly and welcoming man. After chatting for a while about Kelly's research, he then gave thirty minutes of time that probably should have been spent saving the world. Sorry about that.

---

2. Okay, it's a lot more complex than we thought it was even fifty years ago. But with enough computing power and enough smart researchers, we should get a better and better map of the landscape of human maladies and how best to navigate it.

When Dr. Mendelsohn was born, scientists didn't know that our genetic material was made up of DNA. When he was in med school the description in the textbooks for how proteins were made was totally wrong. When he was a young researcher, the amount of data you got from a single experiment was primitive by modern standards.

"If you sequence one human genome you're getting five billion pieces of data. When I began doing my own research, the results could be printed out at first on a piece of paper, and then a long sheet of paper. You can't print out five billion pieces of information and analyze it in your brain. Today at the MD Anderson Cancer Center we are sequencing the DNA of thousands of patients' cancers each year to detect the genetic aberrations that are causing their malignancy."

As medicine becomes more and more personalized, and individual bodies become wellsprings of data, it may be possible to find diagnoses and even cures simply ("simply") by combing through the information for patterns. But we'll only be able to benefit from precision medicine if we collect data from lots of people over time and find ways to store and analyze all of this data, which is going to require a lot of innovation by computer scientists and statisticians.

The National Institutes of Health recently launched the Precision Medicine Initiative Cohort Program. The program is going to collect health and environmental information from over one million participants, including data on their "-omes."

We're going to tell you about a bunch of "-omes" in this chapter, so here's a quick definition—when a scientist adds "-ome" to the end of a word, she means "like . . . all of it." So a geneticist studies particular genes while a genomicist studies all of the genes.[3] Like . . . all of them.

This may make it sound like the genomicist is just smarter, but it's kind of

---

3. This suffix is somewhat useful, though it frequently gets abused because being an "-omicist" sounds really cool. Dr. Jonathan Eisen of University of California, Davis, even keeps a section on his Web site, The Tree of Life, for "badomics," such as the nutrimetabonome, the fermentome, and (for the mystically inclined) the consciousome.

like the difference between a psychologist and a sociologist.[4] The -ome suffix is just a currently fashionable naming convention.

The National Institutes of Health will get these individuals' genomes, microbiomes, and other information, and will then follow them over time to track changes in their health. The data will be available to doctors, who can comb through them for associations among disease, genetics, and environmental factors. This is going to create a massive data set.

Unfortunately, big piles of data don't just leap up and tell you what's going on. This is a problem that generally worries Pfizer's Dr. Sandeep Menon. "The amount of that data that is coming in is pretty much exponential, and . . . the people who are available with the right skill set to analyze is very minuscule. To put it bluntly—the demand is more than the supply."

Dr. Menon is an elite biostatistician. He is vice president and head of the Biostatistics Research and Consulting Center at the largest pharmaceutical company on Earth, and even he finds it "challenging to keep up abreast of the evolving latest techniques to navigate the deluge of data."

What concerns Dr. Menon is that there are a lot of analysts who simply don't know how to handle their data. He thinks a lot of mistakes are being made, which slows the advance of the field and potentially harms patients.

Despite the difficulty of sorting through all this information, progress is being made on many fronts. The field of precision medicine is too enormous to capture entirely, so we've elected to give a bunch of specific examples that should provide you a taste of what precision medicine can do.

Genetic Disorders

The rapid advance in gene-reading technology means that getting your genome sequenced is relatively cheap—in the range of thousands of dollars, where it once costs tens of millions. But being able to diagnose a problem doesn't equal being able to cure it.

---

4. The former is wrong about individuals, the latter is wrong about groups.

Genetic disorders are especially hard to fix. Contrary to how it's often stated in the media, you can't really "edit a person's DNA" the way you edit a document. Almost every cell in your body contains strands of DNA. To alter your genome, you'd have to alter all of your cells, or at least all of your cells that are relevant to your disease.

For example, cystic fibrosis is a genetic disorder where the body produces a lot of thick mucus in internal organs. Thick mucus in the lungs is about as fun as it sounds—it increases the risk of infection while making breathing difficult.

Another problem is mucus in the pancreas. It almost rhymes, but other than that it's terrible—it makes nutrient absorption much harder. Diagnosis with cystic fibrosis actually *does* rhyme, but it once meant that you were unlikely to survive far into your twenties. Advances have gotten patients into their thirties and forties, but these advances are all ways to treat the mucus problem rather than fixing it at its genetic root.

Part of the difficulty is that there isn't just one cystic fibrosis. The mucus buildup that identifies the condition to doctors can be the result of any of a number of genetic mutations.

A drug called ivacaftor was recently developed to deal with a specific genetic variant of cystic fibrosis. The specific mutation that the drug targets is found in only about 5% of cystic fibrosis sufferers. In medicine as usually practiced, that's not great. But in a precision medicine paradigm, the idea is that we eventually have a treatment that targets every possible mutation. So when you are born, we identify your genetic problem and we know exactly what treatment to use. This means you not only get the right treatment, but you avoid a bunch of potentially unpleasant wrong treatments.

But this is still only palliating the problem at a deeper level. What if we could just fix the damaged code in all the right cells?

In the last chapter, we discussed a new gene-editing technique called CRISPR-Cas9. It may allow scientists to actually fix the mutations causing cystic fibrosis in patients with the disease. In case you have already managed to forget what CRISPR does, the short version is that we can now precisely snip out and replace parts of DNA in cells. In principle, this should mean we can snip and fix whatever mutations are causing cystic fibrosis in a living person.

CRISPR has already worked in slabs of intestinal tissue in the lab, which gives you some idea about how fun medical lab work is. Figuring out how to apply the technique in actual patients is still a big hurdle, and scientists are worried that we might mess up other parts of the genome while trying to fix the mutations. When you're editing a trillion cells at once, you don't want to have too many whoopsies.

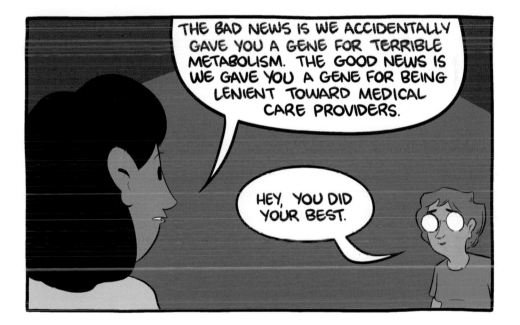

But the cool thing about CRISPR is that it's a general tool for fixing genetic disorders. Any disease that is caused by one or more genetic mutations should be vulnerable to this method of targeted gene edits. If CRISPR ends up being a silver bullet for gene problems, you could fire it at Huntington's disease, sickle cell anemia, Alzheimer's disease, and more.

Cancer Diagnosis, Treatment, and Monitoring

Diagnosis

Cancer cells are hard to kill for the same reason secret androids are going to be hard to kill in the Robot Uprising of 2027: *They look just like us.*

A cancer cell is one of your own cells gone bad. It is a cell that should've served some useful bodily function, but happened to be born with or acquire a weird set of mutations that makes it generate copies of itself over and over instead of doing its job. In the case of a solid tumor, you've basically got a nation of bad cells living and reproducing in your body.

Mutant cells, including cancerous ones, are born in your body all the time. Typically your immune system targets and kills them. The problem is that now and then you get a very rare cell, which (1) reproduces out of control, and (2) evades the immune system, either by avoiding detection or by convincing your immune cells not to kill it.

So by the time you get a cancer diagnosis, you've got very dangerous cells indeed. At that point, medicine has to step in where your immune system has failed. But catching cancer can be hard.

Historically, leukemia is one of the easiest cancers to detect, because it's blood borne and leaves a telltale buildup of white blood cells.[5] Solid tumors, especially small ones, can be much more stealthy. That's why doctors ask patients to do breast exams regularly—even hard tumors can be subtle when they're hidden in a squishy human body.

Early diagnosis is more than just a convenience. Many of the deadliest cancers are dangerous not because they are especially aggressive, but because they only cause subtle symptoms until it's too late to stop them.

According to the National Cancer Institute, there is a 55% chance of surviving lung cancer for five years if you detect it early on. But most people don't get a diagnosis that early. Over half of lung cancer patients aren't diagnosed

---

5. In fact, the word leukemia comes from the Greek root words meaning "white blood."

until the disease has metastasized,[6] at which point the five-year survival rate is about 5%.

So we want to find cancers as early as possible.[7] And it turns out that leukemia isn't the only cancer that leaves a biomarker in blood. All sorts of cancers can be detected by searching for little molecules called microRNA.

In the last chapter, we gave an approximate sense of how DNA creates protein, but we mentioned that the actual process can be more complicated. One layer of complexity is added by microRNA.

MicroRNA is not yet perfectly understood, but one important role it appears to play is in what's called gene expression. Think about it like this: Imagine you have a gene that codes for a glowing red nose. As we said, "codes for" is not a great way to say this, so let's be more specific: Your DNA encodes a protein that automatically goes to the tip of your nose and glows red. How red your nose becomes depends on how many of these proteins are created. MicroRNA can make adjustments here. Suppose your genetic recipe for our imaginary red-nose protein normally ends with "repeat this 10 times." That "10 times" can be adjusted up or down by microRNA, giving you either a pale, sad nose or a ruddy, vibrant nose.

Okay, so that's neat, but so what? Conveniently for medicine, these little microRNA molecules can be found in your bloodstream. Particular microRNA bits, or altered concentrations of them, can tell us not only what cancers you may have, but also what stage those cancers are in.

For example, one study found that levels of four specific microRNAs were a strong predictor of whether someone with lung adenocarcinoma was likely to live a long time (over four years on average) or a short time (a little over nine months). Information like this can help patients and doctors decide how aggres-

---

6. When a cancer has metastasized, bits of the cancer have broken away from the original tumor and have started tumors in new sites.

7. There is some nuance here. Technically, you want to detect cancers as soon as possible, but you also want to know how bad the cancer will be. For example, some prostate cancers grow so slowly that a person is likely to die from natural causes long before the cancer becomes an issue. If you know the cancer isn't going to kill you in your remaining lifespan, then you may choose to avoid the unpleasant therapy, with its risk of impotence or incontinence. Some doctors worry all this data will lead to overtreatment.

sively to approach cancer, and can help patients make decisions about how to live their remaining life spans.

In principle, by knowing what microRNA is in your blood (and maybe a couple of other fluids), we can get a sort of readout on what diseases your body carries. New proteins get created in response to just about anything your body does, so this could be an incredible source of info on exactly what your major malfunction is.

But figuring out what's wrong with you isn't an easy task. Next time you're looking for some reading material, consider miRBase, the microRNA database (mirbase.org), which as of this writing is tracking around 2,000 such molecules.

Another molecule of interest is called ctDNA, short for "circulating tumor DNA." It's a very new discovery, and potentially a very big deal for cancer diagnosis. In simple terms, when you have some types of solid tumor, a little of its DNA can make its way into your bloodstream.

This isn't a great thing for you, but it's very handy for your doctor, for two reasons: (1) It makes it very hard for solid tumors to escape detection, and (2) it means genetic analysis of a known cancer can be done without having to perform invasive surgery.

A recent study showed that if you have stage 1 non-small-cell lung carcinoma, we can already detect it in the blood via ctDNA 50% of the time. By the time you're at stage 2, we can detect it 100% of the time. By this stage, the cancer has moved from the lungs to the lymph nodes, but has not metastasized to other organs. Usually, we don't catch this disease until stage 3 or 4, when five-year-survival probabilities are significantly lower.

But even if you can find ctDNA and microRNA cancer signatures, it can still be hard to get the whole story on your disease.

We tend to talk about cancer in terms of where it's found—liver cancer, bone cancer, brain cancer—but this is not the most useful way to describe it. Two types of breast cancer may come from completely different mutations. One may be deadly while the other is fairly manageable.

Making things more complex is the fact that cancer doesn't stop mutating once it exists. As cancer cells continue to mutate, a sort of survival of the fittest happens in your body. The result can be tumors that are not only dangerous, but also genetically diverse.

The diversity of cancer genetics in a single body can render treatment extremely difficult; even if you have a chemical that makes the tumors shrink, you may be reducing only a subset of the cell types. So if you shrink a tumor, you may have killed only a certain type of cells that were vulnerable. Later, the cancer may come back more aggressive than ever.

Worse, if you receive chemotherapy or radiation therapy, you may have created further mutations in the process. And that's not to mention that going through those treatments often *really* sucks, partially because they are not targeted just to cancer cells. For instance, the traditional chemotherapy treatments target cells that are dividing too fast. But some of your cells, such as stomach lining, are *supposed to* divide fast. It's sort of like cutting down on citywide nefariousness by eliminating everyone with a long black mustache. Sure, you're mostly getting rid of villains, but you're also killing the friendly hipster who makes the best coffee in town. A worthy trade-off? Maybe. But not a pleasant experience.

To defeat cancer, you want to figure out the right treatment early on, so you can avoid giving cancer time to mutate too much. This may mean getting yearly blood tests to detect cancer early, and determine what mutations it carries. This will be important, because it means you can pick the right therapies. Picking the wrong therapy isn't just painful; it's dangerous. You want to know all the relevant mutations so you can make the right cocktail of drugs to treat them.

TREATMENT

One of the reasons cancer cells are so dangerous is that they evade your immune system. It'd be sort of like if there were killer robots on the loose with no conscience, but the ability to mimic humans. But what if we could train the police to recognize these creatures? Like, hey, that one guy who keeps saying

"Affirmative, Human" and works as a lobbyist—maybe we should watch him more closely.

Just so, what if we could teach your immune system to target and kill sneaky cancer cells?

It works like this: You take some blood from your patient, and there you find these immune cells called T cells. The T cells we are interested in are a type that recognize structures on the surface of cells, called antigens. By looking at what type of antigens a cell has, a T cell decides if the cell *needs to die.*

If we know what antigens your cancer cells have, we can teach your T cells to target them.

Why T cells? We'll let Dr. Marcela Maus of Harvard Medical School and Massachusetts General Hospital explain: "There's two things that really make T cells special when it comes to immunotherapy. . . . They have the capacity to kill other cells . . . and . . . they have memory. They're very long-lived cells. And once they've seen something once, they're quicker to recognize it the second time and kill it even faster."

A particularly successful genetic modification has been to teach T cells to go after a molecule called CD19, which is found on a type of white blood cell called a B cell. Leukemia and lymphoma often kill you by overproducing these white blood cells.

One problem with this approach is that the T cells often end up killing *all* the B cells, including the ones that aren't cancerous. B cells are part of the immune system too, so killing all of someone's B cells can make them temporarily immunocompromised. This isn't awesome, but fighting off an infection is generally preferable to fighting blood cancer.

But what about tumors for which killing all of a potential cell type just isn't an option? Defeating a brain tumor by killing all brain cells is a bit of an empty victory.

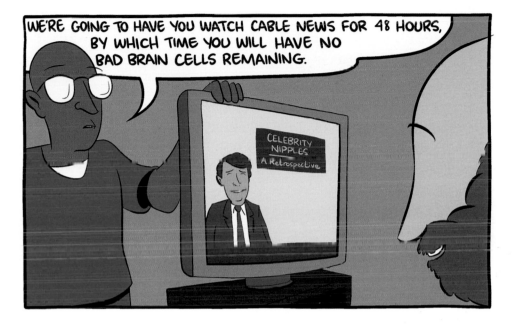

Dr. Maus has a more subtle approach. She engineers the T cells to attack an antigen that goes by the charming name epidermal growth factor receptor variant III, which we'll refer to by the easy-to-remember acronym EGFRvIII. No normal brain cells have EGFRvIII, but *some* tumor cells do. So if you are "lucky" enough to have a tumor with this particular receptor, T cells can be programmed to find and kill them in particular.

Dr. Maus's research is in an early phase, but there is hope that immunotherapy will prove to be a good method for precisely attacking multiple kinds of solid tumors in the future.

MONITORING

If the treatment works, your cancer goes into remission. But at that point you'll still be monitored for the rest of your life. As it happens, being "cured" of cancer is a very good indicator of cancer risk in the future. Precision medicine techniques offer ways to do a better job of monitoring these patients.

For example, if your cancer is in remission, we might still watch for ctDNA, to make sure it's not subtly returning. We might also continually genotype this ctDNA to watch for mutations or a change in prevalence of a certain type. If the ctDNA signature suddenly starts indicating more aggressive types of cancer, it may change the most desirable medical approach.

This new technology is improving rapidly, and we now find ctDNA in stool, urine, and other fluids. It may soon be the case that we can do a much better job of monitoring you for cancer, *and* we can do it by poking through your poop instead of your body.

The Metabolome

One hard thing in medicine is knowing how a particular patient will react to a drug. What works well for this person may work poorly for that person, and some of this is explained by the patient's "metabolome." The metabolome is all of your metabolites: all the small molecules that the machinery of your body operates on, like sugars and vitamins. The metabolome is not a simple system—the Human Metabolome Database (hmdb.ca) is currently tracking about 42,000 different metabolites. That's 42,000 *types* of little molecules. And that doesn't even count the proteins and other "large" molecules that operate on them, making sure you continue to convert cheeseburgers into the energy and muscle required to get more cheeseburgers.

It turns out people have a good deal of metabolomic variation. This may explain why coffee keeps you awake at night, while your friend can have a double espresso in bed, then sleep soundly. You're processing the coffee chemicals in a different way. These differences can tell us about your internal state. For example, we've known for about seventy years that you can tell if someone is diabetic by checking his glucose level. If he isn't metabolizing glucose well, you'll find elevated levels. But with 42,000 metabolites floating around, we might be able to get more information on exactly how different people vary in their metabolic abilities.

For example, some suicidally depressed patients do not respond to any drug regimen. It's possible that something about their metabolome is preventing or altering proper drug uptake. In a precision medicine paradigm, we might be able to determine in advance what the patient will fail to metabolize.

Metabolic information may also help human nutrition. What poisons one person may have no effect on another. This is especially important when the particular poison is tasty. For example, some people have conditions that result in hypocholesterolemia, meaning that for some reason they just don't have as much cholesterol in their blood. This may mean that they can eat more of the delicious high-cholesterol foods that you and I are told to avoid. Maybe there are similar genes that allow people to smoke, or drink to excess, or, like . . . chew glass.[8] Flipping things around, a precision look at a patient may tell you that she should avoid certain foods and activities, even if they're generally considered healthy. Maybe your mom really *was* wrong to make you eat broccoli.[9]

Knowing your metabolome may be especially important for patients who are in a dangerous medical situation. Ideally, any medicine with potential side effects should be given at the smallest possible dose that will work. But if you have a patient with an unusual metabolome, a small dose may act like a large dose, or a large dose may not work at all. If you are choosing among drugs for a patient, knowing how they'll metabolize it may allow you to avoid painful treatments and go straight to the right stuff.

Other Things That Are Trying to Kill You

In the above sections, much of what we've talked about has been cancer and genetic issues. This is because these diseases are some of the most difficult to battle, and we're just now starting to get some major wins. But precision medicine techniques should be applicable to just about anything.

For example, stress and hypertension can cause a condition called cardiac hypertrophy, which is a dangerous thickening of heart muscles. This puts you at risk for things like fatigue, headache, and *suddenly dying*. Given that heart muscles are made of protein, it might not surprise you that there is a microRNA signature for cardiac hypertrophy. In fact, there are a number of microRNA signatures for various types of cardiac hypertrophy. Looking at microRNA in blood gives us a noninvasive way to determine exactly what sort of cardiac hypertrophy you're facing.

---

8. Don't chew glass.
9. She wasn't.

In other words, we might be able to predict that your heart is going to fail before it happens. Given that heart disease is the number one killer worldwide, a bit of a heads-up would be appreciated.

This ability to monitor blood-borne chemicals in depth may also be helpful for patients who are taking dangerous treatments. For instance, suppose you have a patient with a history of heart disease who now needs to take a chemo regimen for cancer. There is an elevated risk of heart failure, and by the time the patient is having a heart attack, it may be too late. With a microRNA analysis, you could monitor the effects of chemotherapy on heart muscles in more or less live time, then make choices accordingly.

Other recent studies have shown that there is a microRNA signature for stroke. There are also microRNA signatures that tell us what your brain is up to as you recover from stroke.

Molecular biomarkers can tell us about other diseases, like Crohn's disease, Alzheimer's, and even what sort of flu you may have. In fact, if you go to Google Scholar and type in "microRNA profile," you'll find an enormous number of papers written just in the past few years, which show that there are signatures for just about every disease from prostate cancer to depression. And not only can we detect many diseases, we can often tell what those diseases are up to.

Some researchers want to know not just the molecular picture or the clinical picture, but also the behavioral picture. Basically, they want the sort of data a creepy stalker would know about you—what TV you're watching, what Web sites you're browsing. This information may clue us in to whether or not you're suffering from some types of diseases. For example, Dr. Andrew Reece of Harvard and Dr. Christopher Danforth of the University of Vermont found they could predict whether an individual was depressed based on the color and brightness of photos the individual posted on Instagram. Instagram photos posted by depressed users tended to have more blues and grays, and generally tended to be darker than photos posted by nondepressed users. Adding this large-scale information to our array of more subtle biomarkers may result in better categorization and treatment for psychiatric conditions.

It may even be the case that this sort of activity monitoring (if done on large groups of people) may find surprising new patterns that correlate with mental disorders. Maybe you think you have anxiety because you're awkward, but actually it's because you are constantly checking how your friends are doing on Facebook.

Social media obsession is a bit easier to diagnose than a slight uptick in a particular bit of genetic code, but it's not necessarily information you know to share with your doctor. There may be all sorts of behaviors individuals and groups engage in that have undetected clinical significance. As wearable computing gets more popular, you may be able to turn in a more full (and more honest) picture to your doctor. Imagine a glorious future where you go to the dentist and have a friendly conversation until a wearable computer betrays the truth about your flossing frequency.

Okay, so there are some privacy issues here, but having an accurate picture of you—from what you're reading and how much you exercise right down to a handful of molecules in your urine—may one day be the way we find what health issues you have and predict what health issues you may acquire.

Concerns

A major problem is and will continue to be cost. Ivacaftor, that cystic fibrosis treatment we mentioned before, will only be used by a few thousand people in the United States. There is probably no economy of scale to be had. So that treatment costs about $300,000 a year. Unless we arrive at extremely generalized cheap methods for drug preparation, it will be hard to drive down the cost of a product with so few consumers.

Another worry is how all this data will affect human psychology. You know your weird hypochondriac uncle? Suppose he now has an encyclopedia's worth of data about his own body and has *proof* that he is suffering on forty-seven metrics. Yeah. Imagine that Awkward Thanksgiving Chat is now *data driven*.

And then there's privacy. To learn more about this issue, we talked to Dr. Kirstin Matthews and Dr. Daniel Wagner at Rice University. We like them. They try to teach ethics to twenty-year-olds, which shows they have a sense of humor.

According to Dr. Matthews, "In order for precision medicine to work and be really useful, you have to connect your genetics with your records, which means all the things that happened to you through your entire life, whether it be injury, environment, or things that happened because of the genetic background. . . . Once that happens there is no anonymity anymore."

So now potential employers or companies selling health or medical insurance can find out who you are and find out whether or not you're likely to suffer from a mental illness or have a debilitating disease. Even if these groups don't have access to health information, they might be able to connect some dots about your mental health simply by analyzing your social media presence. The same group that was able to see depression in your Instagram photo selections was also able to detect depression or PTSD by analyzing Twitter posts. This analysis predicted clinical outcomes months before the official diagnosis was in. Your public behavior may reveal your private health information.

Concepts like life insurance and health insurance only work because it's hard to know in advance who will get sick or die at what time. You don't often think about it, perhaps, but insurance is really a mathematical tool. If a thousand spouses have life insurance, each year some will unexpectedly die. Those people who lose their spouses early receive more money than they put into insurance. Those whose spouses live to old age pay more in than they get out, but they get to keep their spouses longer, which (one hopes) makes up for the loss in cash. In essence, the lucky support the unlucky.

As medicine becomes more and more personalized, this system becomes less and less tenable. There may come a time when all insured people must be genotyped. Those with favorable genotypes will pay less for insurance, while those with unfavorable genotypes will pay more. The lucky will be luckier and the unlucky will be unluckier.

To reduce this risk, the U.S. Congress passed the Genetic Information Nondiscrimination Act of 2008, which made this kind of discrimination illegal. Your employer can't fire you because you have a genetic predisposition toward some medical issue. Nor can an insurer deny you coverage for that reason. But we're all going to have to grapple with what it means to live in a society with extreme medical knowledge. "It's not about protecting the data," says Dr. Wagner, "but about protecting people from the implications of the data." For instance, if you have a genetic predisposition for aggression, are there jobs you should not be allowed to have? And do the people around you have a right to know?

The U.S. government has not made all kinds of genetic discrimination illegal. As an example, take an app called Genetic Access Control, which was made available on GitHub. The app accessed genomic data on 23andMe (a private company through which you get your genome sequenced) and used those data to restrict users' access to Web sites. The app's developer suggests relatively innocuous uses for the app, including creating "safe spaces," like Web sites that can only be accessed by females. But it's easy to imagine how an app like this could be used for more sinister purposes. Imagine a site that only people with a certain skin color can visit, or a site that only individuals lacking genetic defects could visit. Furthermore, even the more benign uses will create problems, because identity is both genetic and cultural. Some people with a traditionally female body type carry XY chromosomes, and a group that genetically barred nonfemales would have to decide how to handle that.

23andMe quickly blocked this app's access to their data, but we can probably expect problems like this to pop up again in the future.

And it's not just *your* personal genetic information. You get half of your genome from your mom and half from your dad.[10] So if you make your genome public, you're sharing half of each of your parent's genomes. In fact, whenever you share genetic information, you are to some degree compromising the anonymity of all people in your kin group. Imagine you have a twin who is in political office, and you find out that you carry a genetic risk of schizophrenia. Do you have a social obligation to share that information? Do you have a filial obligation to hide it?

The potential benefits of precision medicine are vast, but so are the potential costs. As we move inexorably into the era of precision medicine, we should think deeply about what privacy means in an ever more technological society.

But some people believe that privacy matters are less important than the medical secrets waiting to be found in our biological data.

The Personal Genome Project[11] was started by Dr. Church and is a repository for open-access genomic data. Participants are explicitly *not* promised anonymity, and in fact the paperwork on the Web site makes it clear that when personal data are coupled with genomic data, it probably wouldn't be hard for someone to figure out who you are.

---

10. Note that this means acquiring your genomic data may reveal unpleasant truths. For example, you may find out that your biological father is someone other than the man who raised you.
11. The Personal Genome Project is now part of the Open Humans Foundation. Through OpenHumans.org you can upload lots of data (e.g., movement data through FitBit and microbiome data through μBiome) and then share your data with researchers through the site.

The person who told us about the Personal Genome Project is Dr. Steven Keating, whom you may remember from the robotic construction chapter.

During the course of his research on robotic house-building trucks, Dr. Keating discovered that he had a brain tumor. He had surgery to remove it and had his tumor's genome analyzed. For various policy reasons he found accessing the medical and research data on his tumor difficult. He has since become an advocate for making medical data open and available to both the medical community and the patients themselves.

Dr. Keating is excited about the Personal Genome Project. "If you want to contribute your genome, you have to pass a test . . . [that] ensures that you understand the liabilities you're taking. I kind of like that approach, because the whole thing with genetics is that we don't really know what could happen in ten years if you share your data now. You have to go through this list, and you have to sign off on, 'I understand that, yes, a criminal could duplicate parts of my DNA and plant it on a crime scene. Yes, I understand that maybe in the future a virus could be made that's tailored to me; it would only affect me. Yes, I understand that this information could get leaked and somehow get back to my family and show them a condition that they don't want to see.' You have to understand. Then, once you sign off on that, you can submit. . . . It's important that each person understand the liabilities they're taking because it's their data and it's their choice to do so, but the potential benefits in my mind are so much more powerful than those crazy potential issues, and it totally makes sense to me that this will be the future of medicine."

How It Would Change the World

In the medium term, precision medicine techniques will be expensive. In the long term, they have the potential to drive medical costs down by detecting diseases before they become severe, by selecting the right treatments immediately, by curing genetic conditions instead of palliating them, and by tapping into the tendency of computer-related industries to deliver more product for less money over time.

Businesses like Facebook provide users a lot of services in exchange for their data about themselves. Given the value of medical data, a similar bargain may be struck between users and biotech firms. Indeed, Google's parent company Alphabet is already making substantial investments in something called the Baseline Study, which seeks to identify disease-related biomarkers. We don't know their long-term plans, but if the goal is to have a Google Street View of our digestive tracts, we are prepared to pony up whatever data they need.

One of the most exciting effects of precision medicine has to do with how medical trials are conducted. Suppose you had a cancer drug called explodanol. In clinical trials, five sixths of your patients are completely cured, and one sixth of your patients suddenly explode upon taking the drug. This might be a little embarrassing at your next FDA meeting.

But suppose you happen to notice that there is a difference between explodey and nonexplodey patients—the explodey patients all have a particular set of microRNA markers. Now, you can take this drug that was formerly useless because it (seemingly) randomly killed users, and you can successfully cure cancer in a large portion of the population. Also, if you want to explode one sixth of the population, you can now do it in tablet form.

Going further, by getting a genome, microbiome, metabolome, and a bunch of other -omes, you might be able to make very specific patient categories for clinical trials. This is a win on three fronts: (1) It means you can do statistically meaningful drug trials with smaller populations; (2) it means drugs don't have to work for all or even most patients to be considered safe; and (3) it means old drugs that were shelved due to safety issues could be brought back to the market, to be used only on a particular patient category.

For example, it is a little-known fact that there exists a vaccine for Lyme disease. It was pulled off the market because a small subpopulation complained of arthritislike symptoms. So, fun fact: As of 2016, your dog, Mittens, can get a Lyme vaccine, but your son, Mittens, can't.

Although unlikely, it's conceivable that this subpopulation of people who react poorly to the vaccine share a set of characteristics that create the negative side effect. If we could screen out these people, the vast majority of consumers could have access to this important vaccine.

In reality, it'd probably be a bit more complex. That subpopulation may simply have been all the people who are prone to the placebo effect. But given the number of drugs that have failed to get to market over safety concerns, results like this may happen. And with better biostatistical analysis, current drugs could find new uses for difficult diseases. This may seem a little dorky, but it could be enormous. Currently, it costs over $2.5 billion to bring a new drug to market, and most drugs never make it to that point.

One of the books we read on this topic began by discussing "alternative medicine" healing practices. This raised our skeptic hackles, but the author was quickly on to such delightfully *un*natural topics as "Applications of CYP2C19 in Pharmacogenetics" and "Fluorescent in Situ Hybridization." Before he moved on, he made an interesting point—for all their lack of efficacy, one thing those old healers had in their corner was that they claimed to be tailoring treatment to the patient in particular.

The tarot cards had the minor flaw of not actually working, but the approach of saying "What's wrong with you" instead of "What's wrong with *people like you*" must be a deeply appealing one. A person offering a magical cure is willing to say "I know what's wrong and I can fix it," whereas in evidence-based medicine, we can't always know what's wrong, and even if we do it may be hopeless. Precision medicine may give us a way to bring the dreams of the age of magic and make them a reality in the age of science.

10.

Bioprinting

Why Stop at Seven Margaritas When You Can Just Print a New Liver?

Imagine you wake up one day, feeling exhausted and sick. When you finally finish vomiting, you look in the mirror and notice that your eyes and skin have taken on a yellow color. You call your son, who rushes to drive you to the hospital. A few hours later, your doctor enters, looking serious.

You need a new liver.

"Well," you think, "how hard could it be to get a liver? I know tons of people with livers." You look at the nurse. No . . . too small. You look at the doctor. No . . . too old. You look at your son, your eyes glistening. He shakes his head.

"Should've gone to more Little League games, Dad."

So you're stuck. You join over 122,000 people currently on the organ transplant list.[1] Though, lucky you, the line for a liver is merely 15,000 people long.

Depending on your location, health, age, and other factors, you can expect to be on this waiting list for anywhere from a few months to three years. Often the livers go to the people in the worst shape, which means you may need to

---

1. Incidentally, if you live in the United States, we have an opt-in organ donation policy, meaning that your default setting is that if you get killed, your organs die with you. Given that you can't take them with you, wherever you go, as long as it doesn't violate your ethical views we encourage you to opt in.

wait until you go from bad to much worse before you are given priority. And no, you're not allowed to drink your way to getting a liver sooner.[2]

Getting on the transplant list doesn't necessarily mean you'll get a liver in time. Eight thousand people die every year waiting for organs.

But let's say you're a relatively affluent American. Surely there's got to be a better option. You could become a "medical tourist" and go to some other country where you can pay a stranger to "donate" part of her liver to you.

You might want to consider where that liver came from. Or maybe you don't. For example, if you go to China, your liver may have come from an executed prisoner.

Even if you accept the idea of organ sales, and you know the organ was donated voluntarily, well, you may want to consider how you define "voluntary." There is evidence that the most frequent donors in such systems are poor people looking to pay off debt. Often, these people get paid less than they were promised and end up poorer because of lost work time or because a botched surgery affects their health.

So you go back to your family and friends and beg them for a bit of their livers. Livers have a pretty amazing ability to regenerate, so you don't need the whole thing. "Come on," you say. "Just a liiiitle liver."

After a heartfelt display of remorse for missing an important Little League match twenty years ago, you convince your son to donate. Then you discover he's incompatible. HA. So you were right to skip those Little League games. *Apology rescinded.*

If you are able to hold out for a deceased organ donor, you may still have a lifetime of expensive immunosuppressive drugs to look forward to. Since the organ wasn't yours to begin with, your body may recognize it as foreign and sic

---

2. This is actually a serious ethical quandary for hospitals. Given the scarcity of the supply of livers, some hospitals have concerns over treating patients with self-inflicted liver damage. The reasoning is twofold: First, patients who damaged their own bodies could be considered less deserving of help than patients who came by their need "honestly." Second, patients who are alcoholics may be more likely to damage their new organ posttransplant and end up right back on the list. Exactly how to assess these issues is difficult, but many hospitals have a "six-month abstinence" rule for liver transplant patients, to try to winnow out abusers.

your immune system on it. The immunosuppressive drugs may allow the liver to stay, but by suppressing your immune response, they put you at risk of infection.

Then it occurs to you—you've been paying tax dollars into science for years now. Why can't some nerd in a lab coat just science you up a brand new liver? You hear 3D printers are pretty good these days. Can they run off a few copies of a healthy liver for you? You go to Amazon, buy a book on bioprinting, and flip to the preface.

> Only time will tell whether cell printers will truly become organ printers . . . to achieve what the field of tissue engineering has striven for over the last two decades—to create something more than meat-flavored Jell-O!
> • Bradley Ringeisen, ed., *Cell and Organ Printing*

Crud.

Well, okay, better than crud. In fact, researchers are already able to print off usable tissues, and many are working to move the field toward whole organs.

Early attempts to create artificial organs using real cells basically went like this—create a scaffold using some relatively rigid material, then spray it with the appropriate cells. Give the cells something to snack on and watch an organ grow. Sort of like the most horrifying tomato garden ever.

It didn't work very well. The cells weren't dense enough, it took too long to grow, and at least in the early days, the scaffolding material was somewhat toxic.

It turns out that organs are really, really complicated. You can't just get a bunch of liver cells and spray them into a petri dish, any more than you could make a tasty steak by spraying a bunch of "steak cells" onto a plate. Some parts of organs are soft and some are hard. Some stretch, while some are rigid.

In order to understand why it's so difficult to make organs in the lab, we need to talk about *you* for a minute. You may have this idea that your body is made up of a bunch of cells and fluids. This is kind of true, but it misses a universe of stuff. Consider a single newly formed cell. That cell needs energy to move around. It needs chemical signals to tell it where to go. Once it gets where it's going, it likely needs an environmental cue to know what to do. And cells need some structure to hold them in the right place, or else you've now got *runny* meat Jell-O.

Your liver isn't just sitting there waiting to process Happy Hour. It's constantly doing things—sending and receiving chemical information, removing dead cells, getting new ones, and so forth. To make a new one, you need to have the right cells, with the right treatments applied to them, in the right environment, at the right time. It's sort of like trying to create a factory all at once, instead of building the structure, then bringing in machines, then hiring the employees.

We talked to Rice University's Dr. Jordan Miller about recreating this complexity, and he told us: "Although scientists can grow cells in dishes, the amazing structure of the body allows for our rather dense, compact shape. For example, the lung's surface for gas exchange with your bloodstream has the area of a tennis court all folded up and packed inside your chest. So there's simply no way to make a functional replacement for an organ if we can't recreate at least some of that amazing architecture."

Oh, and while you're trying your best to make an organ, the clock is ticking. If a bit of tissue doesn't get a regular supply of nutrients, it can die in a matter of hours. Very thin bits of tissue can absorb nutrients by diffusion (kind of like how a sponge absorbs water), but if you get beyond the thickness of a dime, that doesn't work. You'll end up with a living exterior and a dead core. The Keith Richards of tissue, if you will.

Keeping all that tissue from dying adds a major layer of complexity. In a real human organ, this problem is solved by vasculature—by blood vessels. Blood vessels aren't just the big veins and arteries you're familiar with—they're also the little tiny tubes branching off the ends of larger vessels. These are called capillaries.

Unfortunately, because capillaries are so small, it's really hard to build them onto large blood vessels using current organ-printing technology.

Making things yet harder, organs really need to be perfect. If your heart stops running for two minutes *ever*, you die. Imagine if the whole Internet died every time Comcast had an outage. More important, would you accept a 1% fail rate on your new bladder?

So if you're trying to create new organs, you have a lot of constraints—you need to be able to use many different cell types and give them many different treatments. You need to construct vasculature to supply the organ with nutrients as it is built. You need to do all of this quickly so the cells don't die off while you put things together. And you need to not make any mistakes in the process!

Some researchers think the answer is 3D printing.

There are a lot of ways to 3D print, some of which we described in earlier chapters, but here's the basic idea: By some means, you put down layer after layer of stuff onto a surface. By building up layers on top of each other, you eventually get a three-dimensional object. Generally speaking, this is less efficient than conventional methods of making *a thing*. For instance, if you're making plastic chopsticks, it's cheaper and quicker to inject plastic into a chopstick-shaped mold than to build chopsticks up with layers.

But there are some advantages to 3D printing. For a traditional mold, anytime you want a different shape of chopstick, you have to get an expensive new mold created. A 3D printer can make pretty much any shape of chopstick you want, without any change of machinery. And since you're building up layer by layer, a funky pattern[3] takes about the same amount of time as a more conventional design.

---

3. For we, the undexterous, maybe a funky little bowl shape at the end of one chopstick and a funky little set of four pointed tines at the end of the other.

You can also use 3D printing to interlace different materials. If you've seen a simple desktop 3D printer, typically they only print in one color of plastic at a time. If you add more nozzles, you can get more colors. Industrial 3D printers don't just add colors—they add materials. So you can create unique arrangements of metals, plastics, resins, and other inputs, which would be more or less impossible to do either by hand or using traditional manufacturing methods.

The other cool thing about 3D printing is you can make unusual structures. For instance, suppose you want to create a ball with a honeycomb structure inside. This is impossible with an injection mold, but is relatively easy with 3D printing.

All these qualities make 3D printing a potentially great way to build extremely complex structures, like body parts. In principle, a 3D printer should be able to rapidly shoot out the appropriate cells, proteins, chemicals, treatments, and structural elements needed to construct working human organs.

Better still, a 3D printer would be able to make the product to order for each patient. This is handy, because a five-foot-tall woman does not want the same heart as a seven-foot-tall man.

It's also handy because an on-site 3D printer could use cells harvested and cultured from the individual patient that will receive the organ. For example, you could harvest immature stem cells from someone, coax them into replicating and becoming the cells that make up a particular organ, and then transplant an organ made from these cells back into the patient. Such an organ would not have the rejection problems that normally come from organ transplants.

You may have the impression that 3D printing is mostly for weirdos with too much money who spend their time obsessively printing *Star Wars* miniatures. But this is only about 97% of them. A lot of people are working on projects with a bit more potential, and those of them who are interested in reproducing biological structures have created a field called bioprinting.

There are a lot of ways to do 3D bioprinting. Here are two that are particularly common: Squeeze it or LASER it.

The first technique works essentially the way a frosting gun works. But in this case you're squeezing out bio-ink, which consists of the cells and chemicals that will create your organ, as well as some support goo so you can gently squeeze all this out through the nozzle.

You take a tube full of bio-ink, and you apply pressure to squeeze it through a nozzle. A computer-controlled arm moves the nozzle to the appropriate place at the appropriate time as you build up your layers of bio-ink. This is a pretty good way to go because cells can be relatively delicate and squeezing is relatively gentle. However, any technology that uses extrusion as a method creates problems with flow control and with gumming up the print head. As a heart recipient, you probably don't want to see a medical technician yelling "But I DID cancel that print order, you bastard!" while yanking on a stuck piece of tissue.

Another problem is speed. You're printing cells, which have a tendency to pop when they get squeezed too hard. This limits the amount of pressure you can apply, and thus limits how fast you can go. Going slow is a problem, because cells don't live forever and they may drift after placement.

You can try having fewer cells and more support goo in your bio-ink, but that just means that the stuff you're printing is going to be closer to meat Jell-O than to, well, meat.

Think of it like this—you want to deliver tomatoes to your neighbor at high speed, but your only method is to fire them through a pipe. You can suspend the tomatoes in jelly to protect them, but if you shoot your tomato-in-jelly too hard, the tomatoes will pop when they smack into your neighbor's wall. In fact, just to get them to go that speed, you had to increase the pressure in the tube, which probably already smooshed some of the weaker tomatoes.

It's the same for cells in gels. For systems like this, there is generally a trade-off between the density of cells in your gel and the speed at which your bio-ink can be extruded.

So, how about instead we just use lasers?

LIFT (laser-induced forward transfer) works kind of like how grease fires work. No, really. You know how they say you shouldn't throw water on a grease fire?

When you have a grease fire, you've basically got a pool of oil with its top layer burning. When you toss a cup of water in, two things happen: First, because oil is less dense than water, the flaming oil goes up instead of down, so it doesn't get snuffed out. Second, water boils at 212 degrees Fahrenheit, while oil boils around 600 degrees Fahrenheit. So it's almost like you've got a balloon whose skin is made of flaming oil and whose interior is water that is expanding at an extremely rapid pace. The result is a huge explosion as flaming oil particles fire in every direction, burning down your house, and ruining Pancake Saturday.

LIFT is like that, but on a small scale, and highly controlled. You have a transparent plate, on which you paint your bio-ink. You fire a laser into the back of it. This causes a little vapor bubble to burst in the bio-ink. As a result, a little dot of bio-ink gets ejected and lands on a receiving plate, leaving a gap in the painted bio-ink.

By repeatedly firing the laser this way, you can draw complex patterns of bio-ink dots. By adding more and more layers of these dots, you can build up a 3D shape, as with the frosting gun method. Dots of bio-ink may not seem like an ideal method, but if you can get them to be small enough, and make enough of them, it's sort of like pixels on a screen. All the little bits come together to make a big continuous whole.

To return to our tomato and jelly analogy, it'd be like the tomatoes-in-goo were hanging from a flat surface and you heated parts of the surface to explosively fire tomato goo down. As long as the firing method doesn't pop the tomatoes, you can deliver them just about as fast as you like. Or at least as fast as you can add replacement jelly to the hanging surface.

It may seem like an overly complex way to do things, but it has a lot of virtues—there is no nozzle to get jammed and the results are extremely precise, and you can tune it to have just the right amount of pressure so as not to pop many cells. At that point you can print as fast as your laser can move over the rear surface.

Now that you're familiar with the methods that use bio-ink, we can talk a bit more about what the bio-ink is made of. There are a lot of trade-offs required to get your bio-ink ready for printing in most systems. To understand why, imagine you're trying to make 3D printed cookies.[4] Even if you have a rig for 3D printing cookie glop, you can't just grab some store-bought cookie dough and stick it in your 3D printer frosting gun. The batter might separate, so you have to add an emulsifying chemical. Chunks might get stuck, so you have to remove chocolate chips, nuts, and raisins. The dough might not be homogenous, resulting in ugly cookies, so you have to blend it like crazy before putting it in. By the time you're done, you have cookie batter that extrudes perfectly, but looks and tastes awful. The primary goal of printing a cookie is to get a tasty cookie. But by adding all these other constraints, you make achieving that goal much, much harder.

---

4. This has actually already been done. We haven't had the privilege to try any, but they look okay.

Similarly, when you add stuff to bio-ink to get it to print nicely, you may lessen its ability to form into a proper organ. You'll probably need it to be in a nice gel state so it extrudes well and cushions the cells, so you add alginate. You might want your bio-ink not to evaporate too rapidly, so you add glycerine. You might want it to have totally sweet swirl patterns, so you add red Kool-Aid powder.

Whatever you add, the problem is that none of this stuff is in a real liver. So your additives need to be nontoxic and they need to go away after serving their purpose. And when they do go away, they need to leave behind the right sort of structure.

Some groups have figured out a way to bioprint without some of these bio-ink auxiliary substances and instead rely on the cell's ability to produce a lot of these compounds on their own after they've been printed. However, additives like alginate are still very common.

Even supposing you have the perfect bio-ink, one bio-ink isn't enough. Not nearly. A given organ may have a dozen or more cell types, each of which are differentiated further depending on the purpose they serve.

You'll need different inks. You'll need different *blends* of inks. You'll need different treatments to apply to the bio-ink once it's shot out. For instance, you might want to add a particular chemical, or hit it with UV radiation, or heat it slightly. Or you might want to do all these things in various orders and various intensities. Adding to the headache, you're printing on a wet canvas, and bio-inks may start to bleed. Perhaps literally.

As if all of this wasn't enough, a serious problem in the field is software. 3D printing started in the 1980s. The most common 3D file type is the STL file, which was originally only designed to deal with surfaces of 3D objects. In the case of your new liver, it's probably important to you that there's stuff inside it. Bioprinting scientists have created workarounds and new file types, but there isn't yet a streamlined, agreed-upon framework. Until there is, we're stuck in the '80s.

Okay, so bioprinting is going to be really hard. But the good news is it appears to be a solvable problem. Scientists are already making inroads, and as computers, 3D printers, and our knowledge of organs advances, we get closer and closer to making it work.

Where Are We Now?

At the moment, we're mostly good at printing cell slabs that are about a millimeter thick. We can't get much thicker than this since we haven't perfected printing blood vessels, which means all inputs to and outputs from the cell work through diffusion. But it actually turns out that you can do some amazing things with a thin slab of organ.

Dr. Gabor Forgacs is the scientific founder of Organovo, which (among other things) prints human tissues for drug testing. Being able to test a candidate drug on living human cells before trying it on living humans could save a lot of lives and a lot of money. As Dr. Forgacs told us, "If a drug candidate fails on a—let's say on a piece of human liver, well then, you better think twice before you start administering that drug candidate to patients."

This is a really big deal. Only about one out of every ten drugs make it out of the human clinical trial phase. If you could identify which of the nine drugs will fail before they even get to human trials, then you're saving lives, reducing suffering, and saving millions of dollars on human clinical trials.

There is another reason why it's great to be able to grow thin slabs of human cells. Lots of body parts you might need to have replaced are actually quite thin.

Visualize yourself in the not too distant future walking through an outdoor thin-slab organ market. There's Dr. Caddie Wang. She can make you a brand new cornea! Over here, we have Dr. Jonathan Butcher. Don't mind the name— he's here to make you some heart valves! Down the street, Dr. Gatenholm can set you up with a lovely hunk of cartilage. All these are handy things to have, and all of them are quite thin.

But down at the far end of the market, a few organ experts are trying to give it to you *thick*.[5]

Before you can print thick organs, you need to be able to print blood vessels. Dr. Miller's lab does 3D bioprinting of blood vessels, but in reverse. According to Dr. Miller, "We print the sugar, then we encase in gel. Then we dissolve and remove all sugar to leave a hollow series of connected tubes (it's like the Internet). Then we flow in the vein cells and they stick to the walls of the tubes."

What we're gonna call "the rock-candy-from-hell paradigm" is accomplished using two different methods. The first method they developed uses an open-source 3D printer called a RepRap, with some modifications, including some parts harvested from a toaster. Dr. Miller's toaster saves lives. Ours makes bread more crunchy.

---

5. We didn't realize how weird this metaphor would get when we started.

The RepRap-based model was a frosting-gun-style design, which extruded a specially designed sugar goo, which (for the curious) was indeed edible.

The second method they use is called "sugar sintering."

The word sintering usually refers to an industrial processes where you lay down a layer of powdered metal, then use heat (from, say, a laser) to form it into a solid object. When done in a very precise way, it can be used as a method of 3D printing. You "draw" a shape in the powder with a moving laser, then you add another powder layer and draw again. Like with the frosting-gun method, by building up layers, you arrive at a preprogrammed 3D shape.

In Dr. Miller's case, the powder is sugar. The sugar binds together where the laser blasts it, and after enough layers you've got precise sugar sculptures.

This sugar sintering method has some major benefits when printing the scaffold for blood vessels. Sintering is generally a much more accurate way to create 3D objects than extrusion. This is important when creating the complex structures Dr. Miller is pursuing.

Plus, with the extrusion-based 3D printing methods, support for one layer comes from the layers below it. This makes it difficult to 3D print objects with hangy bits. Like, try to imagine printing a grandfather clock from the bottom up. The lower parts of the pendulum would have to hang in midair until you attached the top.

In the sugar-sintering method, the hanging parts are supported by the surrounding powdered sugar as you build up your layers. Once the structure solidifies and can support itself, the powdered sugar support is simply blown away.

Dr. Miller and his team have been getting some promising results. So far they have printed some of the thicker blood vessels, and the really cool thing is that these blood vessels are acting like they would if they were in a human body. They can withstand the normal pressure that is placed on blood vessels as blood is pumped through them. And they're starting to grow capillaries. On their own.

The Miller Lab is just beginning to explore this technique, and it'll be some time before we live in a world of plentiful livers with nice juicy veins. But it's an important step toward a better tomorrow, where our children can drink twelve beers every night and only worry about the damage done to their friends and family.

Meanwhile, exciting things are starting to happen for simpler body parts. For example, cartilage can be printed a bit thicker than, say, liver tissue because cartilage is weird and relies on slow diffusion rather than blood vessels to get nutrients and remove waste. Dr. Michael McAlpine's group at Princeton 3D printed a "cyborg ear." The fleshy parts of the ear were a combination of cartilage (in this case, made using cells from calves) and silicone, and a coiled antenna was integrated into the ear to allow for a form of hearing.

Okay, sure, an ear is a bit less complex than a heart or a lung. But this method illustrates an important point: Success doesn't require perfect duplication of the natural organ. In fact, organ replication isn't even the most ambitious goal.

According to Dr. Forgacs, "The good news is that there is really no need for us to copy the same structures that are in our body. Who says that our heart is the best machine to pump blood? Or who says that the kidney is the best filter to get rid of poison? Those are highly sophisticated structures, but they carry out a very well-defined function that we as smart bioengineers could possibly mimic. These organs will not be identical copies of the structures that we have, simply because they are not created the same way as the biology does it from early embryonic development, but they will function the same way if not better than our natural organs."

Concerns

We think it's important for us try to locate the dangers of technologies we describe, but frankly, it's really hard to find ethical problems with synthetic organs and their ability to get hundreds of thousands of people off organ waiting lists. To the extent there are concerns, they tend to be specific variants on issues shared by all biotechnology.

For instance, it is likely that the rich will have privileged access to more and better organs, especially at first. But due to medical tourism, this is already true. Right now, the machine that prints the organ is the body of a poor person.

How patent law would work is another issue that is important but not unique to bioprinting. Suppose the Apple iLiver is way better than the Microsoft X-Liver. How long should Apple have the right to keep its product patented?

The various ethical issues of stem cells may arise for organ printing, but at least so far it does not appear to be a major moral conundrum. Most ethical concerns with stem cells are associated with the use of embryonic stem cells. But the stem cells most likely to be used for bioprinters are pluripotent stem cells, which are like embryonic cells but are derived from patients.

A few concerns may be unique to bioprinting, but they seem pretty mild to us. For example, one concern is that printed organs may acquire bacteria during the printing process. In theory, these bacteria could be types that would never get into, say, a liver in the first place. Thus it's possible, though perhaps unlikely, that 3D printed organs could introduce new diseases into the body. But just like sterile surgical technique should keep bacteria out of the body during surgeries, sterile techniques in the bioprinting lab should keep bacteria out of printed organs.

There is a social concern as well—what economists call "moral hazard." The idea is that if you put people in situations where they *can* behave badly, you will probably get bad behavior. The (recently) classic example is a banker who can get a bailout if things go badly for his bank, so he makes stupid loans.

Just the same, if you're not worried about any of your organs, you might start

engaging in more risky behavior in terms of sex, drugs, and cheeseburgers. Perhaps in the distant future, this could be an issue, but it doesn't seem likely to us. Liver surgery isn't exactly fun, and it isn't exactly cheap. And the process of realizing you need a liver transplant isn't typically delightful either.

In short: If bankers had to get their body cavities opened up to receive a bailout, they might think twice about making another risky loan. Your move, Congress.

How It Would Change the World

In the United States, on average, someone dies almost every hour waiting for organs to become available. And while they wait, they are often incapacitated. Bioprinting organs wouldn't just benefit the patient. Society would receive a huge savings in terms of lost work-hours and public medical expenses.

Printed organs would also eliminate the need for organ markets. Your authors are agnostic on whether or not legal organ markets are a good idea[6] (though we're sure *illegal* organ markets are bad news). Legal organ markets are a complicated issue that often balances saving human lives against abuse and exploitation of the poor. Bioprinting could eliminate the need for organ markets, sparing us the ethical conundrum and doing away with the more sinister version of them. Well, assuming there aren't organ elitists who only want their organs "wild caught."

With 3D bioprinting, three-year waiting times to receive a cadaveric organ would be a thing of the past. You would only need to wait as long as it takes to grow yourself a new organ, and this wait time would likely be the same no matter where you live or how healthy you are. Dr. Miller tells us, "It might even be possible to print organ replacements for you way ahead of time, and put them into deep freeze until you need them."

---

6. If this sounds totally crazy, please make sure to read the nota bene to this chapter.

Bioprinting also uses your own cells, so organ rejections and a lifetime of immunosuppressive drugs should no longer be an issue. If it works, not only do you get a new organ, but you have a higher quality of life after it is implanted.

Bioprinting would also provide more realistic subjects for doctors to practice on. This might seem like more of a convenience than a major development, but, well, next time you need surgery you'd probably like for your doctor to have practiced on something more realistic than Mr. Potato Head.

Perhaps, most important of all, it would finally answer that one question we've been dying to have answered:

Nota Bene on Organ-Matching Markets

The way most modern markets work is simple—you see something you want, then you pay money for it. Money is nice because it universalizes goods and labor. When you want to buy a cappuccino, you don't have to offer to clean the barista's apartment or to give her a bushel of carrots.

But in some cases, like romance or organ donation, you can't use money. Or, at least, you can't *just* use money. These situations are called "matching markets," and they come up quite frequently.

One reason we don't use money in these markets is because, well . . . it'd be *weird*. Imagine going to your sweetie, bending a knee, and then saying, "Now, since you're a physical 8 and a social 7, while I'm a physical 5.5 and a social 3, here's a contract for 40 percent of my lifetime earnings, and . . . *Will you marry me?*"

Nope. You can't do that. We encourage you to try, and we sincerely encourage you to videotape the endeavor.

No, you'll just have to find the right match. Hence the term "matching market." This is why love is both a magical experience and an unholy pain in the ass compared to buying a car or a sandwich. Imagine if every time you wanted to buy a meatball sub, you had to write a sonnet to the nice lady slinging cheese at Subway.

> This thou perceiv'st, which makes thy love more strong:
> Toast not the bread, for I must leave ere long.

Historically, which markets are more money driven and which are more match driven has been sociological, with different cultures working it out in their own way. Dr. Alvin Roth (Nobel Prize–winner and author of *Who Gets What—and Why*) at Stanford invokes the concept of "repugnance" to explain why many markets are match driven.

For cultural reasons, and perhaps biological reasons, many transactions are considered to be, well, *repugnant* when money comes into play. Adopting a child is okay. Buying a child is weird. Falling in love is okay. Paying for love is (in modern history) weird. Other issues are in between—paying for sex is acceptable in some cultures and abhorrent in others. In some cultures, health care is something you buy, and in others it's a citizen's right.

In most cultures, exchanging cash for an organ is repugnant. Gifting an organ makes you great. Trading an organ is somewhere in between. That makes it useful, if your goal is to get people off waiting lists.

Dr. Roth studies matching markets, and one of his great successes has been the designing of advanced organ-swapping markets.

To understand how organ-swapping markets work, imagine how transactions might work if there were no money, *but* there were a Web site that could match people who want stuff. Suppose you have a bushel of corn and you want to trade that corn for orthodontia. If there happens to be an orthodontist who wants corn, you're all set.

But if there's not, one option is to introduce a third player. Let's call her Alice. You have something Alice wants, and Alice has something the orthodontist wants: You give corn to Alice. Alice gives surgical gloves to the orthodontist. The orthodontist gives you orthodontia.

This loop of transactions is called a "cycle." Cycles aren't just good because they're pretty. A completed loop means that every person both gave something and received something. In other words, assuming every transaction was voluntary, nobody got screwed.

This may strike you as a bad setup—like, every time you want to buy something, you need to find a cycle of transactions. You're essentially relying on the computer's ability to discover a bizarre coincidence whenever you want something. But if you have enough players in the system and enough computer power, there will be more than enough "coincidences" to make any transaction you like.

Now, imagine you're talking about organs. Suppose you need a kidney. You have a sibling willing to give you one, but it turns out you two are not compatible due to blood type. Meanwhile, two *other* siblings somewhere else are in the same situation. And in a double coincidence, each person needing a kidney is compatible with the other sibling's willing donor!

You agree to a swap. Normally, a stranger doesn't just give up a kidney. But in this case, she's doing it as a way to get a kidney for a sibling. Everyone wins.

(= functional Kidney

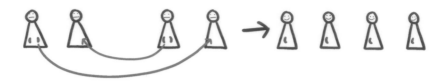

Situations like these are a bit uncommon. But, when you have 120,000 people on the waiting list, you'll find a lot of coincidences.

Organ swapping does open up a potential problem: What if someone changes his mind? Suppose Alice and Andy want to swap kidneys with Barbie and Bill. Alice gets Barbie's kidney on the assumption that Bill will get Andy's kidney. But after Alice gets Barbie's kidney, suddenly Andy's not so keen on giving up his.

This is scary legal ground. You can sue them for money, but can you sue them for a literal pound of flesh?[7] To sidestep this problem, typically organs are swapped simultaneously. All donors get anesthesia at the same time, and the swap happens shortly after.

But this creates a logistical headache, especially for large cycles. If the ideal trade involves, say, five different donors, you have to get five surgical teams working at the same time, and then you have to coordinate the organs around the cycle.

Things like this can, and do, happen. When they do, it's a great thing because five people get taken off the list. But if you can find nice enough people, there's a better way to do things.

To the amazement of your authors, who are as a rule self-interested and narcissistic, there are people willing to give kidneys to strangers for nothing but gratitude. This is not only awesome, but it gets around the reneging problem.

Imagine this. Saintly Sally gives a kidney to Alice because Andy promises to give his kidney to Barbie.

---

7. Okay, about a third of a pound, but still.

But after this is done, Andy turns out to be a kidney-hogging weasel.

Well, it's not a *lovely* situation, but at least Barbie and Bill are no worse off than when they started. Bill still has both kidneys, so they can work on joining another cycle to get Barbie her kidney. Alice is still married to an organ-bogarting jerkface, but at least she's got a kidney.

In actual practice, reneging is more of a theoretical problem than anything. According to Dr. Roth, broken links happen only 2% of the time and "two percent includes not just people who chicken out and changed their mind, but also people who, when they show up, something happens and they can't give a kidney."

The neat thing is that the moment you have a Saintly Sally, you can try to set up a superlong donation chain, where each receiving pair agrees to donate to some other pair.

SALLY ALICE ANDY BARBIE BILL CARL CAIT DON DELIZABETH

In principle, the chain can keep going until someone cheats or something goes wrong. Thus, a bunch of people get kidneys, and they don't need to have simultaneous operations. And from the perspective of an altruistic donor, you're not just giving a kidney—you're starting a whole chain of kidney swaps! The average chain length in the United States is five, but chains of up to seventy pairs have been created. So one Saintly Sally can change a lot of lives.

There are some limitations, though. While people are in need of lots of kinds of organs, matching markets deal almost exclusively in kidneys. This is because kidney exchange surgeries are fairly low risk.

Liver transplant surgery, for example, is a lot more dangerous than kidney transplant surgery. According to Dr. Roth, something goes pretty seriously wrong in about 1 out of 100 liver donations, whereas major complications happen in only 1 out of every 5,000 kidney donations. So while liver exchanges are possible and do happen, kidney exchanges are much more common.

A few readers may be wondering why we don't just set up a cash-for-organs market. Rest assured, all the other readers are mentally glowering at you now. In fact, they think your idea is *repugnant.*

However, the notion of organ-selling markets is well studied, and in fact exists legally in Iran. It also exists *illegally* in a number of places. We don't want to get too into the weeds here, so instead we're just going to propose a thought experiment based on some of Dr. Roth's ideas:

Right now, if you get someone off dialysis via a kidney transplant, the health-care savings (often borne by socialized health care) are approximately $1.25 *million.* This means that a person could be paid $1 million for a kidney and the system would still collect a huge savings. And this doesn't even account for the vastly improved life of the person who gets off dialysis.

Additionally, donors could be given special treatment, like a higher position on the list if they ever need a new kidney. Dr. Roth even proposes special social treatment for such donors, like better airline seating or a badge of honor they could wear, sort of like how veterans are treated today.

Now, suppose that the transaction is also made less repugnant by making it indirect. That is, a rich person can't just point at a poor person and say "Gimme." Organ sellers would be treated just the same as altruistic organ gifters, except they get some money and perks. Their organs would still be given to the people in greatest need according to the metrics used by hospitals.

In other words, an organ market doesn't have to be the organ market of your nightmares. It could be a system where a person can get a lot of money, social esteem, *and* a guaranteed high spot on the donation list, all while saving lives and saving everyone money.

Perhaps by now you're getting a little weirded out. Simply by a little market structuring, and by virtue of the very high payout, the system seems less and less repugnant. One important thing that should always be considered when thinking about markets is whether a new system, however ugly, is better or worse than the current system. A legal money-for-organs clearinghouse might strike you as grotesque, but is it worse than thousands of people dying while on a waiting list?

We don't have an answer for what's right in these markets. In the long run, we hope synthetic organs become possible and then become cheap. In the short run, society will have to do its best to apportion scarce resources as ethically and efficiently as possible.

11.

Brain-Computer Interfaces

Because After Four Billion Years of Evolution You Still Can't Remember Where You Put Your Keys

Brains are all right.

I mean, they're convenient anyway. They're where you put a lot of stuff you like, such as consciousness, memory, identity, and the ability to perceive several lines of black squiggles as a cogent sentence.

But if I asked, "How do you *like* your brain?" your answer wouldn't be entirely positive. You, like a loving spouse who nevertheless has a lifetime of complaints, could list a few dozen things you would like adjusted *immediately*.

Maybe you'd want to be a bit smarter or have better hand-eye coordination. A better memory would be nice. Or, actually, it might be nice to remove a few memories too. How about preventing all bad feelings? And while you're tinkering in there, could you drop in the complete works of Shakespeare? OH! OH! And could you record my dreams? They're really interesting! I promise!

In principle, all these things *could* be done. You know you could be smarter because you know someone who's smarter than you, and you know you could have a better memory because someone else has a better memory. Everything in the brain, from your mental map of your house to the feeling you get while watching a sunset, is physically embedded in your brain. The brain is a physical machine, in many ways similar to a computer. So can we "program" your brain in the same way we program a computer?

Now, we should be careful here. There is a historical tendency to equate the brain with whatever the coolest current technology is. The brain is a clock, the brain is a hydraulic system, the brain is an engine, and so forth.

But there are good reasons to think the brain really *is* something like a computer. Or, more specifically, there are good reasons to think a computer could act like a brain. A clock is a clever mechanism, but it is not a *universal* machine. A clock can only tell time, but a computer—even the piece of crap one in an old flippy phone—can run any program that fits in its memory. Okay, it might take ten years to process the program, but it can run it. If "mind" is in some sense a program, we should be able to run it on a conventional computer.

The 3 pounds of goo between your ears is an amazing machine, with billions of cells and trillions of connections, but let's not forget the virtues of the metal-

and-glass box sitting on your desk. One of the really nice qualities about a conventional computer is that it's easy to change. In particular, you can read from it, upgrade it, and write to it.[1] Scientists already know a way to make changes directly to your brain. It involves opening a book, sitting, and reading.

Nah, just kidding. They're working on ways to make alterations and assessments directly. This isn't just awesome—it might be important. Some people can't communicate or move well because their bodies don't cooperate with their brains. Some people have thought patterns that they'd rather not have. Some of us would like to instantaneously gain knowledge without any effort. Like, wouldn't it be great if you could learn Zen meditation without doing any work on yourself?

A computer can't meditate, and usually doesn't have a body, but it does have some great virtues that a brain lacks. It's designed for humans and by humans, so we know the software and hardware that run it. This means changes are relatively simple. If you want a computer to learn addition, you don't have to spend a year drilling it on two plus two. When you give a computer a new processor, it instantly gets "smarter."

Wouldn't it be nice if we could give that kind of functionality to an actual human brain?

In fact, in a primitive way, we already can. And the methods are improving rapidly. The burgeoning field of brain-computer interfaces is a strange blend of incredibly complex science and rather earthy human needs, in which microelectronics and powerful algorithms let us read brains, restore movement to the paralyzed, restore memory to the elderly, and perhaps give healthy people capacities beyond those of any humans in history.

Also, maybe we can connect a bunch of brains into a big superbrain. Does that sound fun?

---

1. Here we mean "read" and "write" in the broad sense of uploading information and downloading information.

Where Are We Now?

Brain Reading

Reading the brain involves figuring out what thoughts a person is having or what actions that person is engaging in. For example: What words is a person imagining? How is she feeling? Is she thinking about moving her foot? Is she actually moving her foot? It's the most well-studied aspect of brain-computer interfaces, in part because it's comparatively easy. It's sort of like how we're pretty good at observing what a bunch of ants do in a mound (i.e., reading their behavior), but it would be very hard for us to make the ants do something totally new, like spell out the word "HI" (i.e., writing their behavior). If we want to be able to change the brain, figuring out how it operates is the first step.

When trying to interact with a brain, we are somewhat in the position of scientists working with an alien computer. We don't exactly have a USB cord. Even if we did, there's no slot. And even if there was a slot, we don't know how information is encoded. Not exactly, anyway. But we do know the brain emits certain signals, and that those signals correlate to what the mind is doing.

There are two main signal types that we care about—electrical and metabolic.

ELECTRICAL SIGNALS

Your brain is packed with a type of cell called a neuron. These neurons touch each other end to end, and are therefore often referred to as the wiring in the brain. One of the main ways neurons communicate is with electricity. They are able to store a small amount of charge, which they can then discharge (aka "fire") to send a signal to their neighbors. Lots of neurons firing in a certain pattern is what thoughts *are*. When you think, "I'm about to see a drawing of a dancing pie," it's because a particular set of neurons are firing in a particular pattern.

YOUR FEELINGS ARE PURELY PHYSICAL MECHANISMS.

Conveniently, we modern humans have all sorts of contraptions for measuring electrical activity. So we point these devices at the brain and try to figure out what you're thinking from what electrical noise we detect.

METABOLIC SIGNALS

Metabolism is one of those terms you're pretty sure you understand, but if you had to define it you'd fumble for words. In fact, even in science it's a pretty broad term. But basically it means that a chemical is altered in some way that's useful. For example, glucose is a sugar that is converted to a chemical called ATP, which is used to power lots of reactions in your body. Or perhaps you're familiar with ethanol, which is a chemical that can be metabolized into poor life choices.

Because metabolic processes result in chemical change, they leave an effect that we can detect. If ten guys go into a house carrying funny-looking plants and some hydroponic equipment, then those ten guys come out empty-handed, you can *infer* that something's being grown in the basement. Just so, in the brain, when you see oxygenated blood going to one brain region, then deoxygenated blood flowing out, you can guess that brain region is up to something.

Of course, even if we get great signals from the brain, we're still not really *reading* it, exactly. I can pick up a book of Spanish and pronounce it, but that doesn't make me a speaker. If we want to read your brain (and put those signals to use), we need a way to translate. We can do this by correlation.

Suppose, for example, a certain group of neurons lights up with electric signals every time you see a philosophically materialist dancing pie.

THE "SOUL" IS AN UNNECESSARY HYPOTHESIS.

If so, we can safely assume that those groups of neurons have to do with resolving existential angst. And pies. More realistically, if certain groups of neurons light up whenever you move your right arm, we're probably detecting the electric signature of your brain telling your arm to move. This sort of information is particularly useful for creating prosthetic devices. It is also possible to detect more subtle qualities, like whether you're calm or agitated, happy or sad, and focused or scattered.

Now then, let's explore some ways we can already read your brain. As a courtesy to the squeamish reader, we have organized this section from least to most invasive. This should offer you the chance to, as it were, get off the trolley to brain town.

NONINVASIVE ELECTROMAGNETIC BRAIN READING

The classic method for brain reading is the electroencephalogram, or EEG. Basically, you've got a bald cap filled with electrodes, often with a conductive gel. As your brain emits electrical signals, the EEG listens in.

An EEG detects when a whole lot of neurons (let's say 50,000 or so) behave in concert to, for example, lift up your arm. That is, they make some sort of repeated cyclical pattern. Why exactly a huge group of neurons act together (as opposed to a small number of dispersed neurons) is not well understood, but it's a pretty common occurrence in the brain. Conveniently, when you have a lot of neurons acting together, you get a strong enough electric signal that you can detect it outside the brain, using relatively cheap equipment.

EEG has a lot of advantages, the biggest of which is that it doesn't involve any surgery or fancy equipment. Also, EEG gets really good "temporal resolution," as do all electricity-based detectors. You think something, your brain emits a signal, and the EEG detects it, all in a fraction of a second. This is important for any brain-computer interface because the user will want to be able to "talk" to the computer in real time.

The major downside to EEG is that you have poor "spatial resolution." You can detect electrical signals, yes, but it can be hard to tell exactly where they came from. Why? Because you're only detecting signals at the surface of the skull.

Imagine you have a giant sphere filled with a million cats. You have sound detectors all around the rim of the sphere. There is no outside interference because your friends have stopped speaking to you.

When one cat meows, it's too faint to detect. But suppose there's some sort of cat party and a huge group of cats all start meowing at once. Then you detect a signal. Because sound doesn't travel instantaneously, the sound of the cat party will hit some of your detectors sooner than others.

Using your array of detectors, your knowledge of the interior of the sphere, and some computer signal processing, you should be able to tell exactly where the cat party started. But there'll be errors. First off, there are always multiple cat parties at once. And there's a constant background level of meowing. And the meow sounds get a bit distorted moving through the giant sphere and its rim. So instead of being able to say, "The cat party was right here," you have to say, "The cat party started in this general area." Just so, the EEG can tell (in a somewhat blunt way) that "this part of the brain is active." But even getting that data involves some technical hurdles.

For one thing, your face is annoying. Your skull too. Your face uses electrical signals for things like blinking, which are important but not interesting for brain reading. These extra electric signals keep you from getting a perfectly clean EEG reading. Plus, your head can get sweaty or maybe you twitch your scalp a little, which can also mess with the signal.

But given its simplicity and convenience, EEG really is the workhorse of brain-computer interface research. It's cheap. It doesn't require an array of spikes to be pushed through a hole in your head (more on that later), and it does a reasonably good job of detecting brain signals.

One tool that complements EEG is called magnetoencephalography, or MEG. The brain's electric current produces a magnetic field. These magnetic signals are much, much weaker than the electric ones, but they have an advantage—they pass through your skull without much distortion.

In principle, this should mean that you can get really fine signals without getting too invasive. In practice, all current MEG systems are huge. And because of a limitation in the current designs, the device has to sit a little bit away from the skull. This means MEG can't get close enough to get fine spatial resolution.

The real advantage of MEG is that it detects a complementary set of neurons to EEG. See, neuroscientists sometimes model the brain as a cylinder, but this isn't quite accurate. The brain is lumpy. And squishy.

As it happens, this lumpiness means some regions are easier to analyze with data collected by EEG, while others are better with MEG. For analysis purposes, they're better together, but for a brain-computer interface, not so much. MEG requires an extremely sensitive superconducting quantum interference device or SQUID, which we discussed in the introduction. This is why, if you do a Google search for "Meg the Squid," you get information on brain imaging instead of a cartoon squid.

SQUIDs also have to be cooled by liquid helium, which isn't cheap. And the apparatus is so sensitive that the room it's in has to be shielded from the Earth's magnetic field. In short, you probably won't be able to carry it around with you anytime soon.

Noninvasive Metabolic Brain Reading

If you've ever gone to a hospital and been stuck in a narrow tube surrounded by loud noises, you were either in an MRI machine or being born. MRI (short for "magnetic resonance imaging") is a way to electromagnetically "ping" areas of your body, then use the resulting signal to produce an image of the squishy goo inside. For the classic MRI system, you detect differences in the water concentration of tissues, which is useful because tumors tend to contain less water than healthy tissue.

More recently, scientists have created a specialized type of MRI called fMRI. The f is for "functional," which really sounds like the inventor wanted to insult the original MRI. In fact, "functional" refers to how you can watch the brain as it functions. An fMRI takes advantage of the fact that oxygenated blood has a different magnetic signature than does deoxygenated blood. Groups of neurons that are in use consume more oxygen than neurons at rest. By looking for relatively high or low oxygen blood, you can figure out what brain areas appear to be most active. This should allow for some high-resolution brain reading.

The machine is complex, but the logic is simple: If you show me a picture of a cow, and suddenly a particular group of neurons sucks up a bunch of oxygen, those neurons are probably responding to the cow. Or if you ask what my favorite band was in 1993, and a particular group of neurons gets a dose of oxygen, those are probably the neurons that handle shame.

The major downside with fMRI is that it's still a bulky, expensive machine. Like MEG, it's fine for research, but if you want it to be something regular people can have access to all the time, there'll have to be some major technological improvements.

Another metabolic method called fNIRS (functional near-infrared spectroscopy) shows some promise as an alternative.

You know how, when you were a kid, you would take a flashlight and shine it through your hand? A fNIRS is like that, but for neuroscientists.

You take near-infrared light and shine it through a brain. As the light passes through, it is absorbed just a liiiittle differently by oxygenated blood than by deoxygenated blood. A detector on the other side of your head catches the remaining light, and then some fancy computing shows you which brain areas are getting oxygen right now.

Given that detectors and computers are getting ever smaller and cheaper, fNIRS might be a great way to start reading brains for a brain-computer interface. It's arguably even more convenient than EEG, which may require you to put wet gel on your head along with the electrode shower cap.

A big limitation of fNIRS is that the light can't actually get all the way through your brain. In fact, it can only go about an inch. So you have to shoot around the edges. However, loosely speaking, a lot of the stuff we consider active thinking happens at the outer part of the brain, whereas more primitive functions like breathing are governed by inner regions.

The most exciting but least developed metabolic technology is functional magnetic resonance spectroscopy or fMRS (which, sadly, we were informed is not pronounced ffffMissus). The way it works is essentially the same as fMRI, but thanks to relatively recent developments in signal processing technology, we can detect a whole lot more.

As you might imagine, your brain does more than just shuttle oxygen around. In fact, there are thousands of different chemicals with different uses in the brain. So far, fMRS is able to detect some of them. Thus, in addition to telling us *where* stuff is happening, fMRS gives some clue as to *what* is happening, which should provide a far richer picture of what's going on between your ears.

However, fMRS is another technology whose current state-of-the-art equipment is quite bulky and expensive.

And there's one problem we haven't yet mentioned, but it is common to all metabolic devices. Earlier we mentioned that EEG gets good temporal resolution, but bad spatial resolution. That is, it's good with *when* you think something, but bad with *where* the thought occurs in your brain.

Metabolic detectors are the opposite. They give you a really good picture of *where*, but only a marginal picture of when.

Here's a picture of a Nietzschean dancing pie. By the time you finish this sentence, a *metabolic* process pertaining to Nietzschean pies has been completed inside your brain. The corresponding *electrical* process happened before you got to the first letter of that sentence. Metabolic processes usually take about 3–6 seconds, and can take as long as 30 seconds to complete.

BEHOLD, I TEACH YOU THE ÜBERPIE!

This is bad for two reasons: First, for researchers it muddies the data by spreading it out through time. Like, maybe after seeing the dancing pie *that has transcended the morality of the herd,* you also remember a pie your grandma baked. That starts up a different metabolic signal as the first one is winding down, making it hard to tell what's what.

Second, for a brain-computer interface user, it's a big freakin' delay. Imagine you're trying to get a robotic hand to move a bottle of spray cheese toward your mouth. A delay that varies between 3 and 30 seconds is going to be a serious problem, unless you're really into cheesebeards.

Okay, so those are all the common, *nice* ways to read the brain. But what if we could take a more direct approach?

PRETTY DARN INVASIVE BRAIN READING

Frankly, it's annoying to have to just look at the outside of the head and try to make an educated guess about the juicy goo inside. Can't we stick stuff directly on the surface of the brain?

It turns out we can! This method is called electrocorticography, or ECoG. Basically, imagine an EEG but on the surface of your brain instead of the surface of your skull. So it's like an EEG without that annoying hair, skin, bone, and fluid getting in the way.

Guess what? The data is much better than with an EEG. In fact, ECoG has already been used to predict three-dimensional arm movements and allowed paralyzed patients to accurately control computer cursors and even robot arms. The downside is that *there are electrodes on your brain.*

With the modern methods, once the patient heals from surgery, it is possible to make ECoG minimally visible by implanting most of the device under the skin. Just a couple of a wires coming out of the top of your head—nothing you'd be too surprised to see on city streets.

The desires of neuroscientists notwithstanding, ECoG is considered to be only appropriate for patients with serious ailments. So the experiments tend to be done on patients who are having electrodes implanted as a treatment for epilepsy. Several books we read describe people who heroically agree to annoying neuropsychological tests in the intervals between brain surgeries.

While ECoG may have a very important role to play in therapeutic applications, you probably won't be using it to play video games or download your dreams. But many brain-computer interface researchers think ECoG may ultimately be the way to go. It's certainly quite invasive and the procedure still poses serious risk, but once the dangerous phase has passed, it's an excellent source of data that can remain in place for decades at least. As odd as it may seem, ECoG (which, after all, merely sits atop the brain surface) represents a sort of halfway point between neuron-monitoring possibilities.

What's in the other half, you ask?

Really Superinvasive Brain Reading

Suppose you want even better data. There are several options that are more invasive than a suite of electrodes on the surface of your brain, which fall into a category called intracortical neural recording. "Intracortical" is a nice way of saying that stuff is getting stuck into your brain. The classic way to do this is called the Utah Array.

It's not quite as pleasant as the word "Utah" would lead you to believe. What you've got is an array of relatively rigid wires on a square plate. At the tip of each wire is an electrode, which listens for electric brain signals. These wires will probably also have an anti-inflammatory coating before they go in, in order to minimize irritation due to *sticking wires into your brain*.

One nice thing about Utah Arrays is that they improve over time thanks to technical advances. The arrays have gotten better and better, mostly by making the pins more uniform or by increasing their number. It's like with computer chips, but instead it's brain needles.

A similar and somewhat more recent development is called a Michigan Array. It's not terribly dissimilar from sticking a small fork in your brain, though it's a bit more useful. Along the tines of the "fork," you have a series of electrodes that work about the same as those on the needle tips in a Utah Array. In principle, one of the nice things about Michigan Arrays is that they can be made better by adding more electrodes, rather than more spikes. So as the technology improves, you get more data for less damage.

The data from these types of brain readers is particularly good, telling you what electrical stuff is happening at the level of particular neurons or small groups of neurons. The main downside—well, the main downside that isn't the part where your brain gets stabbed—is quality degradation.

Detectors inside the brain tend to rapidly degrade in quality because, as you might guess, the brain doesn't like having metal and silicon stuck in it. When you put your hardware in, the brain mounts an immune response and eventually encapsulates the foreign object in the neural equivalent of scar tissue. After a year or two, you've probably lost signal from over half of your electrodes. It's like having a computer that slowly but surely performs worse, even as you're learning to use it.

One attempt to solve this problem is called a neurotrophic electrode.

Imagine a glass traffic cone, only smaller. About four hundredths of an inch in width. The cone is filled with neurotrophic chemicals (meaning that neurons will grow there).

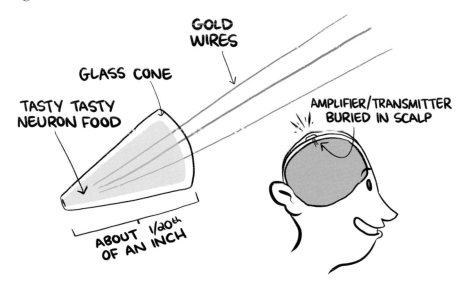

Embedded in the chemicals are electrodes. After implantation in your brain, neurons grow into the gap, where the pins detect them. It's like a little secret garden but with brain spikes.

Given the limited number of electrodes, the amount of information you can get is smaller than what you get with the Utah and Michigan arrays and is quite localized, but it's still quite good. And so far the neurotrophic systems have demonstrated excellent longevity, lasting up to four years in the brains of patients. This particular process is fairly new, so we don't yet know if it will prove as useful as the more established invasive procedures.

In all these inside-the-brain methods, there is a serious issue having to do with long-term tissue damage, beyond the initial damage from insertion. It's like this: Imagine you have a bowl of Jell-O and you stick a fine-tooth comb in it. No matter how carefully you carry around that Jell-O, its wobble is going to make the initial damage from the comb worse over time. In the case of real brain tissue, you get damage and inflammation. These sorts of trade-offs may be reasonable in the case of a person with serious medical issues, but are non-starters if you want some sort of futuristic cyborg brain system.

One approach to fixing things is flexible electrode arrays, so the "pins" can move with the brain. The problem is that if the pins are wiggly, you can't punch them through the Jello-O-ish brain, like you'd like to. So you need a material that goes in hard, but softens when exposed to the moist environment of the brain. And while doing this, it needs to have a tiny electrode tip that sends information back to the surface. Some demonstrations of this sort of thing have already been done in a lab environment, but they aren't yet done in human research.

In all of our brain-reading methods, one general trend you may have noticed is that there's a trade-off between data quality and invasiveness. EEG is relatively comfortable (at least compared to a hole in the head), but the signal-to-noise ratio is fairly bad. The invasive pin arrays give excellent data, but cause a small amount of brain damage. ECoG is somewhere in between. In short, more invasive means better data.

This is a pretty good way to think about brain readers, but there is one other trade-off, between breadth and depth of signal. Pins in the brain get you great data on a tiny part of the brain. EEG gets you rough data on the whole brain. So think about it like this: If you're an alien trying to figure out what's going on with Earth, would you rather have a clear view of a few blocks in New Jersey or an aerial view of the entire planet? From the perspective of a brain-computer interface, it's not entirely clear.

An ideal brain-computer interface should have fine-grain data for the entire brain. There is, in fact, a proposal for this called "neural dust," in which little tiny sensors are all over your brain. We think this sounds amazing, as long as we don't have to try it first.

Upgrading the Brain

Can we upgrade your brain? I mean, your brain has some serious issues. Remember that embarrassing thing you did in high school? *Remember?* Why is it that you remember that, but you can't recall the three laws of thermodynamics?[2]

Okay, so we're not at the point where we can just raise your IQ or give you a better memory or improve your ability to say no to just one more appletini. But, as we discuss below, we may be able to mitigate really serious problems and amplify good things your brain already does well. And we can maybe (*maybe*) improve your ability to learn new skills.

One technology for fixing brain issues is called deep brain stimulation. You can think of this technique as a much more targeted version of electric shock therapy. In a typical system, an electrode is surgically implanted in the brain and connected to a battery placed under the skin.

---

2. GOT YOU. There are four.

Basically, you get a rod put in your brain, and the rod has a battery-powered electrode at its tip. When activated, the electrode sends a continuous high-frequency dose of electricity to the surrounding area. To give an example of why this might be useful, consider someone who is about to have a seizure. Roughly speaking, seizures start in a small region of the brain and move outward from there, in a way that is often compared to the emergence of a storm. Why this happens is not well understood, but by providing a dose of electricity, the deep brain stimulator appears to help stop the storm before it can grow to a full seizure.

This stimulation method is fairly well established for several treatments, since we've been doing it for a while. We spoke with Dr. Aysegul Gunduz at the University of Florida, who explained her experience working with deep brain stimulators this way: "When I started working with the deep brain stimulation group, I realized that *oh my God*, they're actually implanting [deep brain stimulators] and sending these people back home. This is actually approved. The funny thing is we still don't really know how deep brain stimulation improves the symptoms of these diseases."

That's right—most of what we know about how this electric-rod-in-your-head method works is through trial and error. Perhaps, dear reader, your reaction here is "WHAT IN GOD'S NAME? TRIAL AND ERROR?!"

Well, yeah, trial and error is not the ideal empirical method when it comes to sticking an electrode deep in your skull. But there's an issue of ethics here.

Deep brain stimulation isn't something your general practitioner is going to do when you come in with the sniffles. It's typically used on people who have severe problems (e.g., near-constant seizures or suicidal depression) that don't respond to conventional treatments. For these patients, the potential costs of deep brain stimulation are outweighed by the potential benefits. But there is no perfect road map for human brains, so we learn as we go and we typically learn on these patients.

And yet, this method seems to be useful for an ever-broader range of ailments. For example, Dr. Gunduz is attempting to use deep brain stimulation to treat "freezing of gait," a temporary inability to move in Parkinson's patients, and to interrupt and stop tics in patients with Tourette's syndrome.

It might seem a little freaky, but the procedure at this point is fairly common, and it does have some virtues. This simple system appears to have a high longevity, lasting decades in patients.[3] Plus, the more we use this technique, the closer we come to understanding why it works.

One other virtue, at least compared to drugs or psychiatric methods, is that stimulation works more or less immediately after surgery. This is not a small matter for people with dangerous brain conditions. It also makes it easier in principle to know if the procedure actually worked, which is the kind of thing you might be interested in when it comes to voluntary brain shocking.

One company called NeuroPace has created a device called the RNS System that uses ECoG style implants to monitor your brain for the onset of seizures. The device, which is miniaturized enough that you can't tell a patient is using it, delivers targeted pulses when the electrical storm of epilepsy starts.

The RNS System doesn't work for everyone, and it needs a battery replacement at least every five years, which requires brain surgery. Also, unlike the comparatively simple deep brain stimulator, security screening can set off the RNS System. Sudden unwanted electric brain shocks probably make airplane security checks at least 10% worse.

Other researchers are trying a less invasive method called transcranial magnetic stimulation. It's somewhat similar to deep brain stimulation, only it uses powerful magnetic fields, and (this is nice) it doesn't require a hole to be drilled into your head. So far, it's shown to be helpful in pain relief, and possibly in depression. Like deep brain stimulation, it's still a bit of a blunt tool. Scientists identify an area associated with the problem and then throw some magnetic field at it. So far, it seems to have some beneficial results, but studies are ongoing.

But say you don't just want to fix problems—you want to make a healthy brain *better*. That's easy. Exercise, eat right, lower your stress, and study harder.

Nah, just kidding. Can we improve your lazy-ass brain with computers? The answer is *maybe*.

---

3. This may at first seem strange, considering that the more advanced devices we discussed earlier seem to fade in utility in a matter of months. But remember, you're not trying to capture a subtle signal here. You're just firing off electricity. So it doesn't matter if the brain forms a scar around the electrode—it can keep running just fine.

We got in touch with Dr. Eric Leuthardt at Washington University. Dr. Leuthardt is a neurosurgeon *and* a neuroscientist because, you know, you don't wanna put all your eggs in one basket.

He thinks brain-augmenting technologies will follow a similar trajectory to cosmetic surgery.

In the early days of cosmetic surgery, most procedures were aimed at fixing severe disfigurements that were either natural or caused by accidents or war. Once the general public came to accept these methods, their scope was broadened from repair to improvement.

Dr. Leuthardt suggests, "Let's imagine that you get a small implant and you can increase your level of attention. For instance, for something that looks like a small thimble implanted in your skull, you can now enhance your attention and reduce your reaction time. . . . Somebody who's a trader on Wall Street could make an extra hundred million dollars because they're just better at trading for longer periods of time."[4]

For those who are less excited about the brain thimble, there are already studies showing that external magnetic stimulation may improve performance on some cognitive and memory tasks. Perhaps you've heard a headline about people getting magnetic or electric fields run on their brains, and thus having their learning or cognitive abilities improved. This might be true, but a recently released review paper suggests the evidence is mixed.

Along similar lines, very recent studies suggest that deep brain stimulation of areas of the brain associated with memory can enhance learning. These studies have been done on patients who already had the electrode implanted for other reasons.

The evidence suggests that the stimulation improved their ability to remember simple spatial information. It's probably not a good trade-off to get an electrode through your head just so you can remember the way to Burger King, but these results suggest there really *is* something to the idea of using electricity to improve learning. Time will tell.

But even if it's possible, there may be other problems. There is pretty robust evidence that you can study harder if you take amphetamines, but your doctor probably isn't recommending them just yet. The side effects (like increases in blood pressure, blurred vision, and manic symptoms) aren't worth the benefits.

In the case of electromagnetic brain stimulation for brain enhancement, we don't know what the long-term effects are. So you probably won't be able to buy a neurostimulating helmet at Walmart anytime in the next few weeks.

A different approach to "upgrading" your hardware a bit is to simply use what the brain already does, but more efficiently. You know how sometimes you just seem to absorb information faster? At other times, it's like information just bounces off your brain. Sometimes you feel particularly motivated or loving or studious.

The problem is that you may not always know that you're in these states of

---

4. We express no moral qualms about opening the skulls of Wall Street traders.

mind. So perhaps, for whatever reason, you're at an optimal brain state for learning between 1 and 2 P.M. today. But you're spending that time playing video games, so you don't notice. Or you're in a particularly good mental state to exercise between 2 and 3 P.M. today. But you're spending that time playing video games, so you don't notice.

Well, what if these mental states could be detected and brought to your attention? Suppose you had a device that monitored your brain for certain patterns, then told you what you're likely to do well at right this second. You'd know when is a good time to write that poem and when is a good time to fill out spreadsheets. You'd know when is a good time to read Shakespeare and when is a good time to veg out and watch TV.

There is some tantalizing evidence that *if* you detect certain brain waves, it may mean that the brain is in a better state to learn something new. As far as we could tell, however, the evidence that this sort of thing works in humans is quite limited. The most convincing study we found showed that rabbits learned much faster—about two to four times faster!—if you waited to teach them until the right brain wave was detected.

In other words, there appears to be an optimum condition for these rabbits to learn. Might the same hold for humans? So far, we just don't know. Humans, with some exceptions, are a bit more complicated than rabbits. And, in the case of this particular experiment, the learning was an extremely simple motor task.

Writing to the Brain

First off, no, we won't be uploading Shakespeare or calculus or kung fu to your brain any time soon. It turns out this sort of thing is incredibly hard, mostly because memory doesn't work how you think it works. In short, when you experience something, a pattern is created in your neurons. When you recall it, that pattern is, so to speak, replayed. If your brain had been designed by a human for convenient uploading of kung fu, all of your memories would've been stored in a specific chunk of brain, preferably with a USB slot. Inconveniently, nature didn't evolve your skull for peripheral compliance.

Oh, and the other thing is that you already *can* write to the brain, you lazy ass. In fact, you just wrote, "Hey, this book just called me a lazy ass" into your brain. Hearing, smelling, seeing, touching—all the senses are ways to write to the brain.

Well, for most people. For some people (and for many people who are older), these writing mechanisms break down. We're not yet experts on brain writing, but we do know how to use machines to repair some of the old wiring. There are two major technologies that can at least fix broken sight and broken hearing.

You can restore vision using what are called phosphenes. Phosphenes are flashes of what appears to be light that you perceive even though no light has entered your eye. This happens often when people have sudden pressure on their eyeballs. For instance, one of your authors who shall go unnamed really likes to do this thing where he runs up behind his wife, Kelly, and (*gently*) presses on her eyes while shouting "Phosphenes! Phosphenes!" She no doubt enjoys the sudden dance of perceived light before her eyes.

It turns out you can get the same effect via electricity. So scientists have figured out how to implant a device into the eye socket of a blind person that essentially creates an array of pixels using phosphenes. It's not exactly seeing, but it's good enough that someone can get a rough perception of a face. In one case, a person was even able to drive around an empty parking lot.

Hearing can also be restored via a device called a cochlear implant. You may have heard of this, and you may imagine that it's simply some sort of really good hearing aid. In fact, it's a totally different sort of hearing aid. It works like this. A small microphone is placed near your ear. It receives sound, which goes to a receiver placed in your skin. The receiver does its best to filter the sound down to relevant noises (e.g., the voice of the person talking to you rather than the music playing in the background), then it translates the result into electric signals, which are sent through your skull into your inner ear.

By this means, patients who formerly had no hearing are able to hear *something*. Although it takes training, eventually patients are able to hear reasonably well. We've listened to simulations of what cochlear implant patients hear, and it sounds something like a bad cassette tape recording. Which, actually, is pretty damn impressive.

There is one other interesting development in brain writing, which is probably the closest thing going in terms of teaching you instantaneous kung fu. It's called a hippocampal prosthesis, and a few groups are working on it as a way to help with diseases of memory formation, like Alzheimer's and dementia.

When you create a memory, it routes through a part of your brain called the hippocampus, where it can be converted from short-term to long-term memory. When this process is disrupted, making new long-term memories can be hard. This is why your great-grandma remembers details from being a little girl, but completely forgets that today is her birthday.

The idea with this device is that it intercepts the brain signals that are destined for long-term memory, but which would be disrupted by neurodegeneration. The signals are processed, then output to the right place, so that they actually get to be memories.

This may sound like we're actually writing to the brain directly, but here's the catch—we don't know what we're writing. We're just processing it and passing it along. The lead researcher in this area, Dr. Theodore Berger of the University of Southern California, has likened it to translating French to Spanish, without speaking either language.

If you can input memories to long-term storage without "speaking the language," that's pretty convenient, since it potentially means you could record memories from one brain and write them into another. But for the moment, many are skeptical that the process of memory is simple enough for this to be possible.

Concerns

Oh boy, is it easy to have concerns about brain modification.

For starters, the brain is not a simple machine, so making upgrades is not a simple matter. For instance, there is some evidence that mice that are genetically engineered to have better memories are also more susceptible to chronic pain. It's possible that any modification to the way the brain works will have unintended consequences.

And even if there are known undesirable consequences, competition may drive people to brain modification. About one in four elite academic researchers admit to using brain-enhancing drugs. In addition to possibly causing health effects for them individually, this behavior creates a dangerous social dynamic. If one person uses amphetamines to produce work faster, now everyone else is competing with the amphetamine guy for good jobs.

This problem hasn't become profound yet, partially because it's not yet clear that the available brain-modification drugs produce an enormous advantage for users. The lady down the hall who only has to sleep two hours a night is probably more productive than you, but she's not ten times more productive. If new technology starts to create more profound intellectual differences, many people would be more or less obligated to become users.

It's not just high-skilled workers who might be at risk in the brain-computer interface era. Consider this: What right does your employer have to your onboard data? If your implant allows you to be tracked, can the company track you while you're at work? A public-facing company already has some right to ask employees to be in a good mood. If the good mood is modulated at the neural level, does the company have a right to know what the implant is doing?

One suggested beneficial use for employees is a machine that would detect a lapse in on-the-job focus and provide some sort of stimulation. This might increase workplace safety, but we're not sure how we feel about forcing factory employees to have electronic brain focusers.

Having onboard computing power also creates serious privacy concerns. If you have medical equipment in your body, it will likely have some sort of wireless way to communicate with the outside world. This opens up the risk of hacking. In the case of brain implants, hacking could mean a lot of things. For serious implants, a hacker might be able to remotely kill or traumatize an individual. More subtly, by accessing a deep brain stimulator, for instance, a hacker might be able to control your mood or even aspects of your personality.

In general, it might be asked what rights people have to your onboard computing. For instance, if you're taking an SAT and you have a brain interface that lets you talk to the dictionary, do the testers have a right to know? And if you're always going to have the implant, does it matter?

One socially uncomfortable aspect of brain modification is what might happen to current nontypical groups. The blind and deaf have their own communities and, in the case of the deaf, their own language. If neural implants can give people sight and hearing, it might mean the end (or at least the shrinking) of long-standing communities with unique perspectives. In fact, many members of the deaf community have railed against cochlear implants for just this reason.

More ominously, what if behaviors deemed undesirable for cultural reasons could be modified? Okay, we all know a few people who could use some behavior modifying. But consider that this technology will necessarily come into existence at a single point in time, with all of that time period's flaws.

Homosexuality is no longer considered a pathology, but that was not the case for most of history. Indeed, in 1972, Dr. Robert Heath performed experiments on a troubled homosexual man in an attempt to induce heterosexuality using an EEG and electrical stimulation. By altering the brain today, we may extinguish traits we would have later found valuable or morally benign.

This brings us to the most general of concerns when it comes to brain-computer interfaces: Will they be the end of humanity as we know it? If you were to take a healthy newborn from 10,000 years ago and bring her to the present, there is no reason to believe she would have any problems adjusting. The basic hardware of the human brain has been the same for as long as *Homo sapiens* has existed. A brain-computer interface would be the first time we tinker with that. And the first time we try to tinker with the little meaty computer between our ears will necessarily be our most clumsy effort.

Once we gain the ability to modify our own brains, a strange loop is created. We can modify our brains to make ourselves smarter, which means we can make better brain-computer interfaces, which means we can get smarter still, and so on. Pretty soon we're all disembodied ultrabrains with perfect reason, which is kind of a bummer because at that point we won't really enjoy watching *Two and a Half Men* anymore.

How It Would Change the World

As brain-computer interface technology becomes reliable, it may have applications in many industries. An ideal brain-computer interface can make you smarter, better at remembering things, more focused, and perhaps even more creative. You can't think about the possibilities without picturing some sort of cyborg dystopia, and yet, given the opportunity to suddenly have a better memory, we'd probably all take it.

In the short term, most of the utility of brain-computer interfaces will be in fixing problems, rather than enhancing brains. But the neat thing is that fixing broken brain parts requires us to understand them. Understanding should naturally lead to enhancement.

Although you may envision a future where your thoughts are probed by telepathic computers, the major near-term uses for these devices will be therapeutic. One area that is just starting to take off is called neuroprosthetics.

A neuroprosthetic is a device that routes a signal from your brain to a prosthesis. Currently, a lot of the research is on legs because, quite simply, legs are simpler than arms. Neuroprosthetics are seen as the ultimate form of robotic rehabilitation for lost limbs because they do exactly what a real limb does. In very advanced modern prosthetic legs, the machine tries to detect what gait you're trying to perform (walking, jogging, running, skipping) and move accordingly. It's not nearly as natural as a meat-and-bones leg, and it can't perform many of the complex motions you'd want a real leg to be able to do.

Scientists have figured out how to get some of these prosthetics to give the brain feedback. You take it for granted that you don't have to watch your leg move, since you can sense where it is in space and you can feel the muscles in it straining. You can also tell when you're damaging your leg, which isn't possible if your leg can't transmit signals back to the brain. In principle, a brain-computer interface neuroprosthetic should be able to provide some form of back-and-forth communication.

Or if you don't want to build a robot arm, for some patients, you may be able to use a brain-computer interface to make a paralyzed limb move. Paralysis often results from some form of spinal damage. The spine is the main highway for signals from the brain. So when it gets damaged, the signal route from brain to limb gets cut off. A person who can't move her arm due to spinal damage can still send a signal to the arm. The problem is the signal never arrives. But what if we could capture the signal and send it directly to the limb?

A technology called Neurobridge recently did just that. A Utah Array was placed in a patient's motor cortex and wirelessly connected to a series of electrical stimulators on his forearm.[5] With patience and practice, he was able to move his hand and fingers again. The motion is far from perfect, and there is no way for the arm itself to provide feedback, but in the long term, these bridges may represent a cure for limb paralysis.

In addition to the market for advanced therapies, brain-computer interfaces (of a sort) are already being used for more recreational software. A variety of games are available, though they generally use relatively simple technology. A well-titled game called Throw Trucks with Your Mind puts players in a virtual arena where they can win by throwing objects at their opponents. The catch? In order to have the power to throw something, you must exhibit a brain wave that is associated with calmness. Whether this will (as its creator hopes) result in novel therapies for anxiety and attention deficit, or whether it will result in people who are just really relaxed while throwing imaginary trucks, only time can tell.

Another proposal is to improve communication by having computers know your mind state. For instance, a voice recognition device might have trouble telling "I want Kate" from "I want cake!" A simple brain-computer interface might be able to tell whether you're in love or just hungry.

Others long for neuro-cyber-connection. In one set of experiments,[6] lab animals had their brains linked via a brain-computer interface.

---

5. Why his forearm and not his hand? Because the hand is basically a big starfish-shaped puppet you control with your forearm. Go ahead, check it out. Hold up one arm, with your hand slack. Use the other hand to squeeze your forearm, and watch as your slack fingers move.
6. For the record, we're not entirely comfortable with the ethics of this experiment.

The results do appear to be a sort of brain-to-brain connection. We don't know whether these lab rats were literally sharing thoughts, but their connected brains seemed to be able to get together to perform tasks more effectively. At some point in the future, it may be possible to literally combine minds with other people, either for recreational or business purposes.[7] To us, this sounds like a nightmare version of doing a group project,[8] but to each her own.

Dr. Gerwin Schalk of the Wadsworth Center imagines a world where we can communicate experiences more intimately and in more detail. In fact, he thinks it's completely ridiculous that we have to use these stubby meat sticks called fingers to type out our experiences in order to share them. If our brains were connected, you could experience something much more personal. For example, you could put someone in a moment you experienced, and they could see what you saw, smell what you smelled, feel how you felt.

"It would transform society much more so than the computer transformed society. It would completely change essentially what it means to be human. . . . We would be able to interact with technology and with people around us, essentially, just without thinking. . . . It would mean that all our thoughts could readily be aggregated in some cloud. This would completely remove all communication barriers, and society would kind of be some sort of superhuman or something that embodies all the different people. . . . I mean, I can't really think of anything that would have more profound effects on life, really."

We're not entirely sure we're on board for the megabrain thing. We both have lots of thoughts we would prefer to keep to ourselves. Indeed, Dr. Schalk notes that constant connections could have some downfalls: "You sit on the couch with your wife, and you think, 'I want to get divorced from my wife.' All of a sudden, she learns about that. That wouldn't be so great."

No, that wouldn't be so great.

---

7. Or for war purposes. The U.S. Army is also interested in brain-to-brain communication, because soldiers who can communicate wordlessly are less likely to give away their location or plan of attack.
8. "Okay everyone! Let's have a brainstorming session!"

Brain-computer interfaces have this odd characteristic—we want what they can deliver to us individually, but we fear what they would do to us as a society.

Of all the technologies in this book, a perfected brain-computer interface is probably the device with the most unpredictable effects. If a fusion reactor or a space elevator is created, the reactor will still be operated by humans and the elevator will still carry humans to orbit. If your brain is connected to a computer and the two can modify each other, you are not a human being as we have always known ourselves. It would be an end and a beginning.

Nota Bene on Dr. Phil Kennedy

We open this section with a brief transcript from Kelly's interview with Dr. Leuthardt. It actually happens on Skype, but for the sake of drama, please imagine a dark castle room with cold stone walls, littered with strange biological specimens and leather-bound tomes. Add lightning flashes as needed.

> *Dr. Leuthardt:* The technology is not the issue. If I had, actually not that much money, I could build one of those today. . . . It's really the issues of will people finance it, will people develop it, and will the FDA approve it. I think actually in a lot of ways we're much closer than we think. It's really kind of our bureaucracy getting in the way.
>
> *Kelly:* That's fascinating. I didn't realize that we were quite that close to neuroprosthetics for people who don't have disabilities. When you talk to people about getting neuroprosthetics, do you get . . .
>
> *Dr. Leuthardt:* I mean, just as an FYI, this is a fun story. Phil Kennedy . . . he implanted a brain computer. He had a brain-computer interface; he's a normal healthy scientist who implanted a brain-computer interface into himself down in Belize in the past year or two.
>
> *Kelly:* What does it do?
>
> *Dr. Leuthardt:* I think he did it more for scientific reasons so he could study his own brain. It basically records signals from his speech cortex and allows him to control various things. He actually had a problem with it where he had to have it removed, but he did it. He's a normal human being who wanted to have a BCI.

What passes for "normal" may vary between your typical brain surgeon and the average reader, but perhaps that's the price of genius. In any case, the name Dr. Philip Kennedy rang a bell. As it happens, he's the guy who pioneered neurotrophic electrodes, which we discussed in the brain reading section above.

It turns out Dr. Kennedy's story is a bit more complex than the above suggests. He certainly is on the extreme side and, indeed, he flew to Belize to get surgery that wouldn't be approved in the United States, paying $25,000 out of pocket to have an implant put in his motor cortex so that it could talk to a brain-computer interface.

But Dr. Kennedy's decision to get a brain-computer interface wasn't made easily, and it wasn't so that he could become a sort of early cyborg. In fact, his major research goal had been to assist patients with "locked-in syndrome." People with this condition either can't move any part of their body, or can only make minimal motions, like blinking or grunting. If this sounds vaguely familiar, perhaps you've heard of the book *The Diving Bell and the Butterfly*, which was dictated over the course of a year by Jean-Dominique Bauby using only eyeblinks.

A decent amount of brain-computer interface research has focused on ways to give patients like Bauby the ability to communicate more freely. Dr. Kennedy's own work on neurotrophic electrodes allowed a few such patients to operate a computer cursor and select letters. But after about three decades working in this area, the Food and Drug Administration refused to approve more patients for Dr. Kennedy.

Dr. Kennedy was having trouble getting funding and subjects and became worried his life's work was going to be for nothing. It was at this point that he spent his own money to make himself a patient. He wrote out his will, had his company supply him a brain-computer interface, and flew off to Belize.

Two surgeries later, it appeared to be a success. Dr. Kennedy had temporarily lost the ability to speak, which perhaps suggests that the Food and Drug Administration had a point. But he claimed not to feel any anxiety over this. After all, he helped pioneer this surgery, and he was familiar with its side effects.

Maybe he really wasn't anxious about the side effects—it's not just anyone who pays a small fortune to go to a country with looser medical regulations in order to electronically modify his brain.

He conducted extensive research on himself, saying and thinking about words, then monitoring whether the small set of neurons under consideration fired. The data were excellent. Unfortunately, his head never closed up completely, which was, you know, not ideal. After just about a month of research, he had the system removed by surgeons in the United States. And in the most bizarre twist of this story . . . his insurance covered it.

Dr. Kennedy apparently feels it was still worth it. All we can say is hats off to his dedication, and by *God* we wish we could've heard his consultation with the insurance agent.

12.

Conclusion

*Less Soonisher, or
The Graveyard of Lost Chapters*

Before we wrote this book, we were those people who read popular science books and complain about minor inaccuracies—the geeky equivalent of the nacho enthusiast who spits insults from the sidelines of a football match. When the nice people at Penguin accepted our pitch, it was as if the chips-'n'-cheese had been torn from our grasping hands, only to be replaced with, you know, whatever gear actual football players wear.

Nerd pride was on the line.

We did our best to hit the right blend of information and humor, but our biggest fear in writing this book was that someone would call us inaccurate. Or, as we the dorky call it, "the in-word." But given the sheer amount of information we tried to bring together (and the far greater amount of information that had to be condensed or chucked), it's entirely possible we got something somewhere a bit wrong.

So if you happen to notice a factual error please let us know. The best way is to tell your friends and family you're taking a long vacation and will be out of contact, then come to our house and descend the staircase into the dark basement below. There are snacks down there, we promise.

When we first imagined this book, the idea was to take a quick peek at a whole bunch of emergent technologies. Sort of like tapas, only for meganerds. As we went on, we came to feel that we just couldn't bring anything new to the table if we were limited to bite-sized chapters. Honestly, if you just want a thumbnail overview on any of the topics in this book, Wikipedia is a decent place to go. We wanted to bring more depth, more weird specifics, and more of those weird stories you encounter when talking to scientific oddballs or reading obscure documents. Sort of like tapas, only for meganerds of, let's say, larger appetites.

Most of the original topics were excised early on as we lengthened the chapters. Some were rolled together into single chapters on larger topics. But we had a few chapters that we nurtured and cared for, which nevertheless ultimately had to be put down.

By way of concluding the book, we thought we'd let them emerge briefly from the purgatory of a Google Drive folder to enjoy a moment of sunshine before everlasting darkness.

We present: The Graveyard of Lost Chapters.

Grave 1: Space-Based Solar Power

The basic idea with space-based solar power is that you put a gigantic array of solar panels into space and that solar array beams power down to Earth.

There are actually some tantalizing benefits to putting solar panels in space. In space, there's no nighttime, so you don't have to store power or switch to another source after sunset. You can move the panels close to the sun, where you can get more energy per area.[1] You can beam energy down to Earth anywhere you like as long as it has a receiver. And it's environmentally friendly, unless you count millions of tons of rocket fuel that get burned putting all the panels into space.[2]

---

1. If it's not clear why, ask yourself this—does more light hit your face 10 feet from a lightbulb or 1 foot from a lightbulb?
2. Some of the technologies from our chapter on cheap space access would, of course, diminish this issue.

The problem? Well, for starters, it's expensive as hell. A fairly light rooftop solar panel weighs about 20 pounds. At current space launch costs, that's $200,000 *per panel*. Even in the scenario where we have a space elevator and it's a mere $250 per pound, that's still $5,000 *per panel*, not counting the cost of building panels that can function in space.[3] And in the meantime, Earth-made solar panels cost about $200. And the cost is dropping fast.

So why even consider this idea seriously? In *The Case for Space Solar Power*, John Mankins argues that space-based solar panels might get about *forty times* more power per area than ground-based panels. This is because on the ground you have seasons, the day-night cycle, and weather.

This is a somewhat unfair comparison, since space has its own unique problems (like space rocks and radiation), but Mr. Mankins is just trying to establish that space solar *might* be interesting. Fair enough. The problem for us is that, assuming it really is 40 to 1, using our estimates above, the economic benefit of putting solar panels in space is still pretty hard to argue for.

Even if Mr. Mankins's most optimistic numbers are right, *and* we spot you a superadvanced space elevator, at current rates you're paying twenty times the price per panel to get forty times the efficiency. That's fine enough, but we're guessing that by the time we can build a 62,000-mile long space cable, the price of solar panels will have fallen by more than 50%.

So even in the future, when space launch may be quite cheap, it will probably always be cheaper and easier to build forty panels in Arizona than to launch one to space.

It's also easier to do maintenance in Arizona. Okay, Phoenix in the summer isn't *obviously* preferable to the cold void of space, but when it comes to solar power there are a few upsides: Arizona has breathable air, strong gravity, the World's Largest Cow Skull,[4] and a lack of constant high-energy radiation.

---

3. One solution could be to just make the panels in space. Dr. Elvis (whom you'll remember from the asteroid mining chapter) pointed out that lots of the stuff needed to make solar panels can be found in asteroids, so you could gather up the materials from asteroids, make the panels in space, and then transport them over to Earth orbit. This seems like a pretty roundabout way of doing your business, but could one day be possible.
4. In the city of Amado, Arizona. Population 295.

When dirt gets on a panel in Arizona, you fix it with a squeegee instead of a space-robot with a space-squeegee.

Repairing a many-acre solar array in space makes extremely advanced repair robots not only useful but necessary. The alternative is to have a large crew of astronauts, trained to withstand the rigors of the star-speckled sea of blackness beyond the pale of sky . . . just so they can wash windows all day.

Maybe we'll get the right sort of robots via some of the developments discussed in this book, but *even if you had an autonomous swarm of repair bots*, you'd rather have them repairing panels in Arizona, home of the World's Largest Petrified Tree.[5]

One possible virtue of solar arrays was that they might (in the very distant future) be valuable for space transport. The idea is that you harvest energy from panels near the sun and then beam power to vehicles that are already in space. This might actually be the way things go in the long term, since the sun represents an enormous amount of free energy. But by the time we can launch megastructures with robot repairmen onboard, we're guessing there'll be better options.

We looked at a lot of far-out technologies for this book, and no doubt some of them will not materialize, or at least not materialize in the form we find most exciting. But space solar seemed to us to be undesirable even under its most ideal conditions. Perhaps if our energy needs become far, far greater and we decide only to use renewables, we could one day literally run out of usable real estate. But this seems unlikely. At current solar panel efficiency, the entire world's power needs could be met by covering less than 10% of the Sahara Desert with a solar power plant.[6] We're guessing people would appreciate the shade.

In its defense, a lot of smart people think it has possibilities. We remain skeptical.

---

5. In the city of Holbrook, Arizona. Population 5,053.
6. For the nitpicky: We are ignoring the losses due to transmitting Sahara-based power to, say, Canada. That said, in reality solar panels are spread out all over the place. We're just being a little dramatic to make our point.

Grave 2: Advanced Prosthetics

What are advanced prosthetics? It's a bit of a broad category, but really we were interested in anything that takes us beyond the paradigm of solid hunks of wood or metal, toward a world where prosthetic limbs are functionally closer to old-fashioned meat-based limbs. That covers a lot of cool stuff, from advanced materials that are strong yet flexible, to limbs with onboard computers that try to anticipate your motion, to the neuroprosthetics we touched on in the last chapter.

Modern, noncomputerized prosthetics are already pretty amazing. There are advanced materials and designs so that even devices that appear to be quite simple can be both functional and beautiful. And, 3D printers certainly have a place, both for making custom-fitted prosthetics and for adding intricate designs to the usually dull prosthetics available in mass manufacture.

Computerized prosthetics are also quite neat. As we mentioned, some artificial legs try to determine exactly what gait you're doing—jogging, running, skipping, stair climbing, moonwalking—and respond accordingly. They aren't perfect, but it's a big improvement over the old solid-hunk-of-wood method.

One of the downsides to computerized versions is that they are basically another device that needs to be charged. This is obvious to anyone using a prosthesis, but it was kind of a revelation to us. If you have a smart limb, every night you have to take it off, stick it on the charger, and feel irritation at people who complain about having to charge their iPads.

The difficulties of improving prosthetics are fascinating. For instance, the actual motion of the human leg is more complex than you realize. Part of why a peg leg sucks is that it only has one level of stiffness. Think about it. For any gait you choose (assuming you have two standard-issue meat legs), you automatically put your leg at a certain level of stiffness. Try walking with your knees locked. Sure, it looks *supercool*, but it's slow and painful. Now, try walking without engaging your knees at all. Or maybe don't. Or if you do, don't sue us when you break your face.

There are in-between stiffness regimes that are specific to walking, to running, to jumping, and so on. You learned all this by the time you were four years old, so you just don't think about it, but it's kind of incredible. When you walk, your knee and ankle conspire to give your leg just the right amount of springiness so you don't hurt your joints and so your leg (like a giant spring) recaptures some of the energy spent moving forward. Meat-based legs also provide feedback, so that even when you get it wrong as an adult, you make rapid adjustments. These kinds of feedbacks are being integrated into prosthetics, but the technology is still very rudimentary.

And then there are ankles. You have really, really nice ankles. For instance, when you run, you don't keep them straight. You actually flick them out to the side just a little. The faster you're going, the more pronounced the flick. The lowly ankle is actually a pretty neat little contraption, which can move along any axis and stiffen in place at all sorts of positions. Try to imagine soccer without being able to move your ankle any way but forward and back. Even bowling would be tough. Worst of all, without ankles, Victorian gentlemen would have nothing to ogle.

And we haven't even mentioned hands. Hands are the ultimate prosthetic challenge because there is just way too much stuff going on with them. Just to point at something, you make decisions with every knuckle of your hand, plus your wrist, and probably your elbow and shoulder.

Hands also pose a problem from a powered-motion perspective. If you lose your hand (say, your dad cuts it off in a duel), you'll want a replacement. As we mentioned in the last chapter, the muscles that operate your fingers are really in your forearm.

If you lose your hand *and* forearm, mechanical muscles can later be put in the prosthetic forearm to operate the prosthetic hand. If *you just* lose the hand, those mechanical muscles have to be inside the hand itself. That doesn't leave a lot of space to sneak in little batteries and actuators.

This is why when Darth Vader cut off Luke Skywalker's hand, he was being even more of a jerk than the casual viewer might notice. By cutting off his *hand* but not his *forearm*, he left Luke needing a much more complex prosthetic. Such is the power of the Dark Side.

The future of prosthetics is probably in neuroprosthetics, which directly route signals from brain to appendage. As we also discussed in the last chapter, we're getting there, but it's still a ways off from working really well without being extremely invasive. Someday, we may yet have replacement limbs that work as naturally as real ones, but given the history of prosthetics, small improvements would mean a much better life for a lot of people. And in the really awesome future, maybe we'll get extra limbs or brand-new kinds of limbs. Tentacles for everyone!

So why'd we kill this chapter? Partially, we worried that the more technical descriptions of prosthetic challenges would get a bit repetitive and dull. It was fascinating in a dorky way, but we figured most readers don't want to hear all about the degrees of freedom of a typical ankle joint.

More important, a lot of the most exciting developments in this field are really redundant with the stuff in the brain-computer interface chapter. Neuroprosthetic technology is its own research area, but it's really a subfield of brain-computer interface stuff.

So, not without some regret, we consigned it to the grave.

Here lies
Advanced
Prosthetics

Ashes to to ashes
Dust to dust
Silicon to silicon-dioxide polymorph alpha
Iron to rust

Grave 3: Room-Temperature Superconductors

One of the really exciting things about this chapter was the opportunity to go beyond the one thing people (perhaps) know about superconductors—that they conduct without loss of energy.

It turns out that lossless power transmission probably isn't as awesome as we guessed. We got some numbers on how much power you'd save if you replaced all current wiring with lossless transmission, and it was around 10% or less. That much more power is nothing to sniff at, but consider that—even if you had a superconductor that required no cooling—to get the full 10% back, you'd have to remove and replace every bit of transmission wire currently in use.

Plus, all the high-temperature superconducting materials we know of are not in a handy form. If you saw one in the wild, you'd probably think it was some weird sort of rock. Copper, on the other hand, can easily be stretched into nice long wires. If at some point we do discover a superconductor that can operate at human-friendly temperatures, we're betting it's not going to be as easy to work with as a hunk of bendable metal.

So transmission is probably out.[7] But it's not even close to the only game in

---

7. One strong counterargument to our point here is that lossless transmission would mean any power plant could power any receiver, since the length of the line isn't a problem. This could be especially good for renewables, which often are in isolated areas, like deserts.

town with superconductors. These weird objects have two other amazing properties: the Meissner effect, and flux pinning.

This is the chapter graveyard, so we can't give you all the details, but in short, the Meissner effect is this: When an object cools down enough to go from a nonsuperconductor state to a superconductor state, it rapidly expels magnetic fields from its interior. So, for example, imagine you sat a regular magnet on an uncooled superconductor, then started taking the temperature down and down and down. Eventually, you get the superconductor beyond the critical temperature at which it starts superconducting. Suddenly, the magnet leaps away!

Why? Because the magnet has magnetic fields and the superconductor is Meissnering[8] them away.

Next, there's flux pinning. In short, for certain types of superconductors (called Type II), the Meissner effect is kind of half-assed. These superconductors allow magnetic fields to penetrate them here and there instead of being expelled entirely. This weird quality means you can "pin" a magnet to a superconductor.

When you combine these two effects, things get really weird. You have a magnet and a superconductor. You cool the system down. The magnet *tries* to jump away, but it can't because it's flux pinned. The result is that it just . . . floats there, where you pinned it. It'd be like if you leashed yourself to the celebrity you find sexiest. They're constantly trying to get away, but the leash holds them in place. So they just stay at the exact same distance.

This is called superconducting levitation.[9]

The magnet does not merely float above the superconductor. It is pinned in place. If you turn either object sideways, they'll maintain the same separation. You can even turn them upside down. It's as if they are repelling each other, yet connected by invisible strings.[10]

The potential applications of this property are many. For one thing, you now

---

8. Not actually a real word.
9. The magnet floating thing, not the thing where we try to convince John Oliver that he loves us.
10. Please, John. Give us a chance.

have a connector that never wears out because the two connected objects never actually touch.

It also means you can spin the floating magnet and it will never stop spinning because there is no friction.[11] This motion can later be harvested for power. So levitation becomes an incredibly good way to store energy.

Levitation is also a way to have trains that go extremely fast. These magnetic levitation trains (which you may remember from the chapter on cheap access to space) are already being deployed in some places, though they remain quite expensive. The fact that the train never actually has to touch the track cuts down on wear and tear, and means you don't lose any speed to friction between the wheels and the track.

Room-temperature superconductors would potentially cut the price of operating these trains way down, given that you wouldn't need to use coolant liquids or refrigeration to keep the magnets cold. Or if you *did* use those things just to be safe, you'd be able to use less.

Right now, and for the foreseeable future, you have to use liquid coolants to make superconductors work. Since the 1980s, we've gotten "high-temperature superconductors" to work (at a balmy -300 degrees Fahrenheit or so), which means we can use relatively cheap coolants. Very recently, that's been improved to about -100, which is a temperature you can get with a fancy fridge or by going to Antarctica at the right time. However, the superconductor that worked at that temperature needed to be at a pressure comparable to the bottom of an ocean trench.

Also, fun fact, the material used was hydrogen sulfide. Sulfide, as in sulfur—the chemical element that smells like rotten eggs. So the world's warmest superconductor is probably also its smelliest.

Ultimately, we decided to bury this chapter for two reasons: First, the scientific explanations we wanted to provide really required quantum mechanics to get right. And the nonrigorous explanations we *did* have were still pretty hard

---

11. There will actually be a very small amount of friction unless the surrounding area is a perfect vacuum.

on the brain. Second, when we interviewed people, the general impression we got was that the field was pretty pessimistic about widespread adoption of room-temperature superconductors, if and when they are discovered. We interviewed Dr. Inna Vishik at the University of California, Davis, and she said that although the applications for a new material are often hard to predict, advancing the temperature of superconductors is likely more valuable for research science than for, you know, cool new stuff.[12]

It's a shame, because we had this especially good nota bene about a researcher who famously said he was working on ytterbium barium copper oxide, when he was actually working on yttrium barium copper oxide. It was *wild*.

Grave 4: Quantum Computing

Oh, quantum computing. Rise for a moment from your coffin.

How you nearly broke us.

This chapter, despite taking more of our research time than any other, ulti-

---

12. Incidentally, adding to our blossoming understanding that all scientists, even ones working on macroscopic quantum effects, are entirely human, she mentioned that in the forgettable 1989 Richard Pryor/Gene Wilder buddy comedy *See No Evil, Hear No Evil*, there's a twist at the end involving a coin that is actually a room-temperature superconductor. She apologized for the spoiler. Too little, too late, Dr. Vishik.

mately had to be scrapped. The really terrible part is that we almost can't tell you why. That's the tough thing with quantum computing—find an article on it, and that article is probably wrong, or at least simplified beyond the possibility of being right.

Usually quantum computing is treated like a sort of next phase in computing, or as a way to get your computer a magic speed boost, or as a means to access infinite universes (also called the multiverse or parallel universes) to do infinite computations at once. These ideas are related to the truth, but are far enough from it to be misleading.

To really understand quantum computing (and we just barely got to where we sorta kinda maybe vaguely got it), you really need to understand what a computer is doing at the level of bits, *and* you need to have a decent knowledge of basic quantum mechanics. And even if you understand these things, it's not immediately obvious how they come together to solve math problems.

So, writing this chapter we found ourselves in the position of trying to explain how an inanimate sequence of on and off switches can run a video game or play a song or operate an eerily human-seeming chatbot (really, it's weird, isn't it?), while also explaining all sorts of quantum oddities, like generalized rules of probability involving negative and even complex numbers. We had this tortured explanation about building a Schrödinger's cat setup, then daisy-chaining it to more Schrödinger's cats, where if cat A was alive cat B was dead, unless cat C . . . you know . . . et cetera. After writing about two thirds of the chapter, it was already far longer than the others, and that was before we added what we optimistically call "humor."

Thus, unlike Schrödinger's cat, this chapter definitely became dead.

The shame of it is that we absolutely fell in love with this field of study. When quantum computing is publicly discussed, it's mostly because a quantum computer could crack the most common method of digital data encryption. Quantum computing has other applications, like certain forms of database searching or calculating the behavior of atomic-scale objects, which would be very important for research science.

But it also might have implications for our understanding of what existence is. This is possible because the way it works actually involves many universes (or at least it sure looks like it does).

As Dr. Scott Aaronson,[13] a major figure in the current field, told us what's compelling about quantum computing is that it requires you to *really actually* accept what quantum mechanics tells us. All that stuff you may have read in a popular science treatment of quantum mechanics about particles in two places at once and a thing not being determined until you measure it—it's not just for theorizing or amusement. In a quantum computer, that bizarre stuff is the real guts of a machine that produces results that you can print off on an inkjet and hold in your hands.

Indeed, Dr. David Deutsch, the founder of the field, believes that the existence[14] of algorithms that can only run on quantum computers in our universe *proves* the existence of infinite universes. He issues a challenge to other scientists in his book *The Fabric of Reality*, asking them to explain the successful performance of a particular method of number factorization that is probably impossible for traditional computers to do for large numbers: "So, if the visible universe were the extent of physical reality, physical reality would not even remotely contain the resources required to factorize such a large number. Who did factorize it, then? How, and where, was the computation performed?"

A device for solving math and physics problems that could, by the mere virtue of its existence, have implications for our understanding of the entire cosmos? That's as beautiful as it gets.

---

13. Dr. Aaronson was kind enough to have several long conversations with us. His blog, *Shtetl-Optimized,* is a delight. Another man who gave us his time was Dr. Jonathan Dowling, who wrote an incredible and somewhat underappreciated book called *Schrödinger's Killer App.* Both of these researchers are (against all justice in the multiverse) delightful people, brilliant scientists, and incredible writers.
14. There aren't really practical quantum computers yet, but there are some very small setups that actually do work! That said, the results themselves are significantly less interesting than the fact that they were achieved using quantum bits. So far we can use quantum computing to prove that the prime factors of 21 are 3 and 7.

Leaving the Graveyard

If you are young and reading this book, many of these proposed revolutions may happen in your lifetime. That means you can be a part of them if you're willing to work for it. Most of the people we talked to in this book aren't famous—they're working academics like Kelly or deep, probing thinkers, also like Kelly. Contact them! On any given day, your typical academic is a mildly lonely person working in a gray office. Their love can be bought with cookies. Cheap ones. That second sleeve of Thin Mints on your minifridge could be what gets you to Mars in 2050.

We hope that, unlike so many books, we have not tried to sell you on a philosophy of futurology, or on a vision of the future. To our way of thinking, it's probably impossible and it's certainly not necessary. It's exciting enough to know that right this second, people far smarter than us are working out how to probe your thoughts one neuron at a time or to pry open distant alien minerals.

L. P. Hartley wrote in *The Go-Between* that "the past is a foreign country." If that's true, the future is a foreign country too. We're in a little, landlocked nation called The Present, and as far as we may think we can see, eventually the curve of the future turns away and down, leaving us only the narrow band of horizon.

But what a horizon!

Nota Bene on Mirror Humans

Okay, the book's over. Nobody's looking. Let's talk about *mirror humans*.

Oh, you haven't heard of mirror humans? Let's back up a moment.

Life is made of lots of little molecules, and these molecules make up important bigger molecules, like DNA, RNA, and proteins. Some molecules exhibit what's called chirality, from the Greek word for "hand." If a molecule has chirality, that basically means there is a mirror version of it.

To wrap your head around "mirror versions" of things, think about your hands. They look exactly the same, but no matter how you rotate your left hand, it won't be exactly the same as the right. If you have your palms up, your left thumb will point left and your right thumb will point right. Each hand has all the same parts, but they are flipped, as if through a mirror.

When you have two molecules that are mirror images of one another, one of these molecules is designated as the left-handed version and the other as the right-handed version.[15] Intriguingly, life seems to favor a certain handedness for particular tasks. For example, almost all amino acids (which you may remember from previous chapters are the building blocks of proteins) are in the left-handed form. Why nature abhors right-handed amino acids is a topic of debate, but even the amino acids we find in space tend to be left-handed.

But, screw nature. Whatever her reasoning is, there is no known *physical* reason that we couldn't create an organism out of completely opposite-handed molecules in the lab—a "mirror organism," if you like. Some scientists, including Dr. Church, are working to create (simple) mirror organisms, with the hope of one day creating larger and larger such creatures. Why exactly would we want this? Well, for one, it's awesome. You create something that looks like a nice little kitty, but is totally incompatible with the rest of the life on the planet, perhaps even the universe. For example, mirror-opposite organisms would need

---

15. Whether a molecule is designed as the left-handed or right-handed version depends on how polarized light rotates as it passes through the molecule.

to eat mirror food in order to be able to digest it. They would also be undigestible to all predators. Best of all,[16] a mirror-opposite organism would be completely immune to all diseases, because all living parasites and pathogens evolved to infect organisms with normal chirality.

And if it worked, hey, we could scale up to making mirror-opposite humans. Mirror humans would be immune to all the diseases that have plagued humanity for centuries. Malaria? No problem. Tuberculosis? Meh.

Okay, so there'd be downsides too. They'd need mirror food, perhaps mirror microbes. And, if a mirror disease evolved, they'd need mirror medicine. They'd also need mirror partners if they wanted children.

Ah yes, what of love between mirror-original and mirror-opposite humans? Couples comprised of different mirror-type people would, you know, fit together. But they would not produce any living offspring, because you can't mix left-handed and right-handed people when it comes to genetic material. Look, we're not chirality bigots. We think heterochiral couples might get along just fine, but we just worry about the children. Mostly because they wouldn't get to exist.

Although mirror-opposite people would look pretty much the same to our eyes, they would be members of a separate species. Being genetically isolated populations, we would slowly go from being similar but incompatible to having different physical and psychological characteristics over time. Given that we mirror originals would be comparatively disease riddled, it probably wouldn't be long before the mirror-opposites looked at us like we were a shambling horde of zombies.

Speaking of which, what is this chapter doing here in the graveyard? In an earlier version of this book, mirror organisms were going to be a subsection in our chapter on synthetic biology. After some reading, we were a little confused and we had some doubts about the utility of mirror organisms. Making entirely new types of beings just so they wouldn't get sick seems like a rather circuitous path to wellness. And it was our sense that this idea isn't so much a scientific

---

16. Because one of us is a parasitologist.

field as a really neat idea batted around by a small number of synthetic bio nerds. Mirror microbes might some day have research utility—for instance, you could study mirror smallpox without a risk of it getting loose to infect people—but, even that possibility is probably a very long ways away.

But if mirror organisms are a no-go, mirror molecules might still be amazing. Like, what if you could make a tasty sugar that you couldn't actually metabolize? A scientist named Gilbert Levin had this idea in the 1980s and actually discovered a tasty "mirror sugar" that could be used as a no-calorie sweetener. Unfortunately, it turned out to be so expensive that the process never caught on. In a related bit of gustatory history, a product that mimicked oil called Olestra was put on the market in the 1990s. Olestra made nice crispy potato chips without all those fat calories. However, one downside was that some eaters of Olestra products experienced an increased rate of (squeamish readers beware) "anal oil leakage." This being one of the less good leakages, Olestra has largely been shelved. We don't know what the side effects of eating mirror sugar would be, and perhaps that's for the best.

One less disgusting fact we wondered about was whether you could detect mirror humans based on how they responded to mirror versions of flavor molecules. It just so happens that the molecule that gives caraway seeds—the distinct flavor in a Jewish rye—their carawayish taste is a perfect mirror of the molecule that gives spearmint its spearminty taste. We wanted to know if a mirror human would think Jewish rye tasted like some unholy sort of spearmint bread.

To find out, we talked to Dr. Steven Munger, who is the director of the Center for Smell and Taste at the University of Florida. First, he politely pointed out that we were asking the wrong question. The question isn't whether caraway or spearmint would *taste* different, it's whether or not they would smell different. "Taste happens in the mouth and is limited to things that elicit sensations of sweet, sour, bitter, salty, and umami (savory) . . . and maybe fat. Flavor combines taste and smell. For many spices . . . the major contribution of the flavor is the smell." So we learned something that, in retrospect, we were surprised we hadn't realized earlier.

But we still needed to know—can we use tasty rye bread to see if mirror people walk among us? It turns out it's hard to say. In fact, it's hard even to think about. For mirror humans to smell the mirror molecule, their smell receptors would have to be mirrored such that they would bind to spearmint molecules and send an "I smell caraway!" message to the brain. This is possible, but it's not obvious that it would work this way in practice. As Dr. Munger told us (probably as he wondered how the hell he got into this conversation), "At the end of the day, this is really in the realm of guesswork."

So basically what we're saying is, if you have a friend who thinks rye bread is gross, we can't be 100% certain that Dr. Church didn't built him in a secret lab.

Acknowledgments

A surprising diversity of scientists, medical doctors, and engineers took time away from making the world a better place to talk to us. This probably wasn't a good idea for them or for, you know, society, but we really appreciate it. Many of these experts were kind enough to read their relevant sections of the book, or (in a few cases) the whole damn thing. We thank Aysegul Gunduz, Gerwin Schalk, Eric Leuthardt, Beth Shapiro, George Church, Joff Silberg, Pamela Silver, Ramon Gonzalez, Marcela Maus, Steven Keating, Kirstin Matthews, Daniel Wagner, John Mendelsohn, Sandeep Menon, Jordan Miller, Gabor Forgacs, Alvin Roth, Erik Demaine, Cynthia Sung, Skylar Tibbits, Serena Booth, Alan Craig, Caitlin Fisher, Gaia Dempsey, Jonathan Ventura, Justin Werfel, Kirstin Petersen, Christopher Willis, Behrokh Khoshnevis, Richard Hull, Daniel Brunner, Bruce Lipschultz, Alex Wellerstein, Robert Kolasinski, Margaret Harding, Per Peterson, Jessica Lovering, Jason Derleth, Ron Turner, Michel van Pelt, Phil Plait, Daniel Faber, James Hansen, Martin Elvis, Karen Daniels, Steven Munger, Bryan Caplan, Noah Smith, Inna Vishik, Kevin Ringelman, John Timmer, Jonathan Dowling, Alan Winfield, Andrew Reece, Jeffrey Lipton, David White, Aindrila Mukhopadhyay, Sridhar Ramesh, Gerhard Schall, Nick Matteo, Cin-Ty Lee, Dana Glass, Omar Renteria, Javier Omar Garcia, Greg Lieberman, Brian Pickard, Michael Johnson, Scott Egan, Scott Solomon, Paul Robinette, Patricia Smith, Martin Weiner, Alexander Roederer, Rick Karnesky, Rhett Allain, Alexander Bolonkin, Lloyd James, James Lloyd, Ann Chang, Sean Leonard, Scott Aaronson, Rosemary Mosco, Aaron Sabolch, Joe Batwinis, Emily Lakdawalla, Steven Cavins, Jacob Stump, Linda Novitski, James Ashby, Ian McNab, Jennifer Drummond, James Cropcho, Daniela Rus, Kurt Schwenk, Chad Jones, James Redfearn, Kevin Berry, and Richard Prenzlow. Any mistakes in the book are entirely our fault. Or, wait, no. They're Phil Plait's fault. Yeah.

We also wish to thank all of our followers on Twitter and Facebook who helped us find leads when we were stymied and who helped us understand basic concepts in unfamiliar fields. There are too many of you for us to thank individually, but thank you. If you wondered why we were asking so many oddly specific questions, well, you're holding the answer.

We are particularly thankful to our editor, Virginia Smith Younce, for helping us to evolve this book to a form we are very proud of. We thank our copy editor, Jane Cavolina, for making us look far, *far* smarter than we actually are. We thank our production editor, Megan Gerrity, for dealing with our pathetic technical difficulties. And we thank Penguin counsel Karen Mayer, and we wish her well in seeking vengeance on her family.

We wish to thank our representatives, Mark Saffian of Content House and Seth Fishman of the Gernert Company. Life is so much easier when the people representing you are friends.

Kelly would also like to thank Grounds for Thought Coffee Shop in Bowling Green, Ohio, for being a particularly excellent place to drink coffee and write a book.

Finally, we are thankful to our daughter, Ada, who was remarkably happy and joyous even when this book made us less attentive than we would like to have been. We would've loved her even if she was yelling and screaming the whole time, but her mellow presence and big, toothy smile helped us remember why we were doing all of this. And if she some day remembers 2016, it will be "the days of cartoons and takeout." We love you, peanut kid.

If anyone who helped us was not included in the above, we apologize.

Bibliography

Aaronson, Scott. *Quantum Computing Since Democritus.* Cambridge: Cambridge University Press, 2013.

———. *The Scott Aaronson Blog.* "Shtetl-Optimized." 2016. scottaaronson.com/blog.

Adams, James. *Bull's Eye: The Assassination and Life of Supergun Inventor Gerald Bull.* New York: Crown, 1992.

Akhtar, Allana. "Holocaust Museum, Auschwitz Want Pokémon Go Hunts Out." *USA Today,* July 12, 2016. usatoday.com/story/tech/news/2016/07/12/holocaust-museum-auschwitz-want-pokmon-go-hunts-stop-pokmon/86991010.

Alloul, H., et al. "*La Supraconductivité dans Tous Ses États.*" supraconductivite.fr/fr/index.php.

American Society of Plastic Surgeons. "History of Plastic Surgery." 2016. plasticsurgery.org/news/history-of-plastic-surgery.html.

APMEX. "Platinum Prices." Apmex.com, 2016. apmex.com/spotprices/platinum-price.

Artemiades, P., ed. *Neuro-Robotics: From Brain Machine Interfaces to Rehabilitation Robotics.* New York: Springer, 2014.

Autor, David H. "Why Are There Still So Many Jobs? The History and Future of Workplace Automation." *Journal of Economic Perspectives* 29 (2015):3–30.

Babb, Greg. "Augmented Reality Can Increase Productivity." *Area Blog,* 2015. thearea.org/augmented-reality-can-increase-productivity.

Badescu, Viorel. *Asteroids: Prospective Energy and Material Resources.* Heidelberg and New York: Springer, 2013.

Ball, Philip. "Make Your Own World With Programmable Matter." *IEEE Spectrum,* May 27, 2014. spectrum.ieee.org/robotics/robotics-hardware/make-your-own-world-with-programmable-matter.

Baran, G. R., Kiani, M. F., and Samuel, S. P. *Healthcare and Biomedical Technology in the 21st Century: An Introduction for Non-Science Majors.* New York: Springer, 2013.

Barfield, Woodrow. *Fundamentals of Wearable Computers and Augmented Reality, Second Edition.* Boca Raton, Fla.: CRC Press, 2015.

Barr, Alistair. "Google's New Moonshot Project: The Human Body." *Wall Street Journal,* July 27, 2014. wsj.com/articles/google-to-collect-data-to-define-healthy-human-1406246214.

Bedau, Mark A., and Parke, Emily C. *The Ethics of Protocells: Moral and Social Implications of Creating Life in the Laboratory.* Cambridge, Mass: MIT Press, 2009.

Bentley, Matthew A. *Spaceplanes: From Airport to Spaceport.* New York: Springer, 2009.

Berger, T. W., Song, D., Chan, R. H. M., Marmarelis, V. Z., LaCoss, J., Wills, J., Hampson, R. E., Deadwyler, S. A., and Granacki, J. J. "A Hippocampal Cognitive Prosthesis: Multi-Input, Multi-Output Nonlinear Modeling and VLSI Implementation." *IEEE Transactions on Neural Systems and Rehabilitation Engineering* 20 (2012): 198–211.

Bernholz, Peter, and Kugler, Peter. "The Price Revolution in the 16th Century: Empirical Results from a Structural Vectorautoregression Model." Faculty of Business and Economics, University of Basel. Working paper. 2007. https://ideas.repec.org/p/bsl/wpaper/2007-12.html.

Bettegowda, C., Sausen, M., Leary, R. J., Kinde, I., Wang, Y., Agrawal, N., Bartlett, B. R., Wang, H., Luber, B., Alani, R. M., et al. "Detection of Circulating Tumor DNA in Early- and Late-Stage Human Malignancies." *Science Translational Medicine* 6, no. 224 (2014):224ra24.

Blundell, Stephen J. *Superconductivity: A Very Short Introduction.* Oxford and New York: Oxford University Press, 2009.

Boeke, J. D., Church, G., Hessel, A., Kelley, N. J., Arkin, A., Cai, Y., Carlson, R., Chakravarti, A., Cornish, V. W., Holt, L., et al. The Genome Project-Write. *Science* 353, no. 6295 (2016):126–27.

Bolonkin, Alexander. *Non-Rocket Space Launch and Flight.* Amsterdam and Oxford: Elsevier Science, 2006.

Bonnefon, J.-F., Shariff, A., and Rahwan, I. "The Social Dilemma of Autonomous Vehicles." *Science* 352, no. 6293 (2016): 1573–76.

BIBLIOGRAPHY

Bornholt, J., Lopez, R., Carmean, D. M., Ceze, L., Seelig, G., and Strauss, K. "A DNA-Based Archival Storage System." *Proceedings of the Twenty-First International Conference on Architectural Support* (2016):637–49.

Bostrom, Nick, and Cirkovic, Milan M. *Global Catastrophic Risks.* Oxford and New York: Oxford University Press, 2011.

Botella, C., Bretón-López, J., Quero, S., Baños, R., and García-Palacios, A. "Treating Cockroach Phobia with Augmented Reality." *Behavior Therapy* 41, no. 3 (2010):401–13.

Boyle, Rebecca. "Atomic Gardens, the Biotechnology of the Past, Can Teach Lessons About the Future of Farming." *Popular Science*, April 22, 2011. popsci.com/technology/article/2011-04/atomic-gardens-biotechnology-past-can-teach-lessons-about-future-farming.

Brell-Cokcan, S., Braumann, J., and Willette, A. *Robotic Fabrication in Architecture, Art and Design 2014.* New York: Springer, 2016.

Brentjens, R. J., Davila, M. L., Riviere, I., Park, J., Wang, X., Cowell, L. G., Bartido, S., Stefanski, J., Taylor, C., Olszewska, M., et al. "CD19-Targeted T Cells Rapidly Induce Molecular Remissions in Adults with Chemotherapy-Refractory Acute Lymphoblastic Leukemia." *Science Translational Medicine* 5, no. 177 (2013):177ra38.

British Medical Association. "Boosting Your Brainpower: Ethical Aspects of Cognitive Enhancements. A Discussion Paper from the British Medical Association." Discussion paper. London: British Medical Association, 2007. repository.library.georgetown.edu/handle/10822/511709.

Broad, William J. "Useful Mutants, Bred with Radiation." *New York Times,* August 28, 2007. nytimes.com/2007/08/28/science/28crop.html.

Brown, Julian. *Minds, Machines, and the Multiverse: The Quest for the Quantum Computer.* New York: Simon and Schuster, 2000.

Buck, Joshua. "Grants Awarded for Technologies That Could Transform Space Exploration." NASA press release. August 14, 2015. nasa.gov/press-release/nasa-awards-grants-for-technologies-that-could-transform-space-exploration.

Callaway, Ewen. "UK Scientists Gain Licence to Edit Genes in Human Embryos." *Nature* 530, no. 7588 (2016):18.

Campbell, T. A., Tibbits, S., and Garrett, B. "The Next Wave: 4D Printing—Programming the Material World." Atlantic Council. May 2014. atlanticcouncil.org/images/publications/The_Next_Wave_4D_Printing_Programming_the_Material_World.pdf.

Carini, C., Menon S. M., and Chang, M. *Clinical and Statistical Considerations in Personalized Medicine.* London: Chapman and Hall/CRC, 2014.

Carney, Scott. *The Red Market: On the Trail of the World's Organ Brokers, Bone Thieves, Blood Farmers, and Child Traffickers.* New York: William Morrow, 2011.

Centers for Disease Control and Prevention. "CDC Media Statement on Newly Discovered Smallpox Specimens." CDC News Releases. July 8, 2014. cdc.gov/media/releases/2014/s0708-nih.html.

Chandler, Michele. "Alphabet, Apple in Tech Health Care 'Convergence.'" *Investor's Business Daily,* January 21, 2016. investors.com/news/technology/alphabet-google-looking-to-health-care-for-new-medical-products.

Cheng, H.-Y., Masiello, C. A., Bennett, G. N., and Silberg, J. J. "Volatile Gas Production by Methyl Halide Transferase: An In Situ Reporter of Microbial Gene Expression in Soil." *Environmental Science & Technology* 50, no. 16 (2016):8750–59.

Cho, Adrian. "Cost Skyrockets for United States' Share of ITER Fusion Project." American Association for the Advancement of Science. *Science,* April 10, 2014. sciencemag.org/news/2014/04/cost-skyrockets-united-states-share-iter-fusion-project.

Church, George M., and Regis, Ed. *Regenesis: How Synthetic Biology Will Reinvent Nature and Ourselves.* New York: Basic Books, 2014.

Clayton, T. A., Baker, D., Lindon, J. C., Everett, J. R., and Nicholson, J. K. "Pharmacometabonomic Identification of a Significant Host-Microbiome Metabolic Interaction Affecting Human Drug Metabolism." *Proceedings of the National Academy of Sciences* 106, no. 34 (2009):14728–33.

Clery, Daniel. *A Piece of the Sun: The Quest for Fusion Energy.* New York: Overlook Press, 2013.

Clish, Clary B. "Metabolomics: An Emerging but Powerful Tool for Precision Medicine." *Cold Spring Harbor Molecular Case Studies* 1, no. 1 (2015).

Cohen, D. L., Lipton, J. L., Cutler, M., Coulter, D., Vesco, A., and Lipson, H. "Hydrocolloid Printing: A Novel Platform for Customized Food Production." Paper presented at the Solid Freeform Fabrication Symposium, Austin, Texas, August 2009. 3–5.

Cohen, Jean-Louis, and Moeller, G. Martin. *Liquid Stone: New Architecture in Concrete.* New York: Princeton Architectural Press, 2006.

Cohen, Jon. "Brain Implants Could Restore the Ability to Form Memories." *MIT Technology Review* (2013). technologyreview.com/s/513681/memory-implants.

BIBLIOGRAPHY

Colemeadow, J., Joyce, H., and Turcanu, V. "Precise Treatment of Cystic Fibrosis—Current Treatments and Perspectives for Using CRISPR." *Expert Review of Precision Medicine and Drug Development* 1, no. 2 (2016):169–80.

Complete Anatomy Lab. "The Future of Medical Learning." Project Esper. 2016. completeanatomy.3d4medical.com/esper.php.

Computer History Museum. "Timeline of Computer History: Memory & Storage." 2016. computerhistory.org/timeline/memory-storage.

Conant, M. A., and Lane, B. "Secondary Syphilis Misdiagnosed As Infectious Mononucleosis." *California Medicine* 109, no. 6 (1968): 462–64.

Cong, L., Ran, F.A., Cox, D., Lin, S., Barretto, R., Habib, N., Hsu, P.D., Wu, X., Jiang, W., Marraffini, L.A., et al. "Multiplex Genome Engineering Using CRISPR/Cas Systems." *Science* 339, no. 6161 (2013): 819–23.

Construction Robotics. Home of the Semi-Automated Mason. 2016. construction-robotics.com.

Contour Crafting. "Space Colonies." Contour Crafting Robotic Construction System. 2014. contourcrafting.org/space-colonies/.

Cooper, D. K. C. "A Brief History of Cross-Species Organ Transplantation." *Proceedings of Baylor University Medical Center* 25, no. 1 (2012): 49–57.

Craig, Alan B. *Understanding Augmented Reality: Concepts and Applications.* Amsterdam: Morgan Kaufmann, 2013.

Craven, B. A., Paterson, E. G., and Settles, G. S. "The Fluid Dynamics of Canine Olfaction: Unique Nasal Airflow Patterns As an Explanation of Macrosmia." *Journal of the Royal Society Interface* 7, no. 47 (2010):933–43.

Cullis, Pieter. *The Personalized Medicine Revolution: How Diagnosing and Treating Disease Are About to Change Forever.* Vancouver: Greystone Books, 2015.

Daniels, K.E. "Rubble-Pile Near Earth Objects: Insights from Granular Physics." In *Asteroids*, edited by V. Badescu, 271–86. Berlin and Heidelberg: Springer, 2013.

DAQRI. "Smart Helmet." 2016. http://daqri.com/home/product/daqri-smart-helmet.

Delaney, K., and Massey, T. R. "New Device Allows Brain to Bypass Spinal Cord, Move Paralyzed Limbs." Battelle Memorial Institute press releases. 2014. www.battelle.org/newsroom/press-releases/new-device-allows-brain-to-bypass-spinal-cord-move-paralyzed-limbs.

Delp, M. D., Charvat, J. M., Limoli, C. L., Globus, R. K., and Ghosh, P. "Apollo Lunar Astronauts Show Higher Cardiovascular Disease Mortality: Possible Deep Space Radiation Effects on the Vascular Endothelium." *Scientific Reports* 6 (2016): 29901.

Department of Health and Human Services. "Becoming a Donor." 2011. www.organdonor.gov/becomingdonor/index.html.
———. "The Drug Development Process—Step 3. Clinical Research." Food and Drug Administration. 2016. www.fda.gov/ForPatients/Approvals/Drugs/ucm405622.htm.

———. (2011). "Fiscal Year 2015: Food and Drug Administration, Justification of Estimates for Appropriations Committees." Silver Spring, Maryland: Food and Drug Administration, 2015. www.fda.gov/downloads/AboutFDA/ReportsManualsForms/Reports/BudgetReports/UCM388309.pdf.

Department of Labor, Bureau of Labor Statistics. "Industry Employment and Output Projections to 2024." Monthly Labor Review. December 2015. www.bls.gov/opub/mlr/2015/article/industry-employment-and-output-projections-to-2024.htm.
———. "Industries at a Glance: Construction: NAICS 23." 2016. www.bls.gov/iag/tgs/iag23.htm#fatalities_injuries_and_illnesses.

Department of State. "Treaty on Principles Governing the Activities of States in the Exploration and Use of Outer Space, Including the Moon and Other Celestial Bodies." Bureau of Arms Control, Verification, and Compliance. 2004. www.state.gov/r/pa/ei/rls/dos/3797.htm.

de Selding, Peter B. "SpaceX Says Reusable Stage Could Cut Prices 30 Percent, Plans November Falcon Heavy Debut." Space News. March 10, 2106. spacenews.com/spacex-says-reusable-stage-could-cut-prices-by-30-plans-first-falcon-heavy-in-november.

Deutsch, David. *The Fabric of Reality: The Science of Parallel Universes—and Its Implications.* New York: Penguin Books, 1998.

Donaldson, H., Doubleday, R., Hefferman, S., Klondar, E., and Tummarello, K. "Are Talking Heads Blowing Hot Air? An Analysis of the Accuracy of Forecasts in the Political Media." Hamilton College paper, Public Policy 501. 2011. hamilton.edu/documents/An-Analysis-of-the-Accuracy-of-Forecasts-in-the-Political-Media.pdf.

Dondorp, A. M., Nosten, F., Yi, P., Das, D., Phyo, A. P., Tarning, J., Lwin, K. M., Ariey, F., Hanpithakpong, W., Lee, S. J., et al. "Artemisinin Resistance in Plasmodium *falciparum* Malaria." *New England Journal of Medicine* 361, no. 5 (2009):455–67.

Doursat, R., Sayama, H., and Michel, O. *Morphogenetic Engineering: Toward Programmable Complex Systems.* Heidelberg and New York: Springer, 2012.

Dowling, Jonathan P. *Schrödinger's Killer App: Race to Build the World's First Quantum Computer.* Boca Raton, Fla.: CRC Press, 2013.

BIBLIOGRAPHY

Driscoll, C.A., Macdonald, D.W., and O'Brien, S.J. "From Wild Animals to Domestic Pets, an Evolutionary View of Domestication." *Proceedings of the National Academy of Sciences* 106, supplement 1 (2009):9971–78.

Drummond, Katie. "Darpa's Creepy Robo-Blob Learns to Crawl." *Wired,* December 2, 2011. wired.com/2011/12/darpa-chembot.

Duan, B., Hockaday, L. A., Kang, K. H., and Butcher, J.T. "3D Bioprinting of Heterogeneous Aortic Valve Conduits with Alginate/Gelatin Hydrogels." *Journal of Biomedical Materials Research* 101A, no. 5 (2013):1255–64.

Dunn, Nick. *Digital Fabrication in Architecture.* London: Laurence King Publishing, 2012.

Eccles, R. "A Role for the Nasal Cycle in Respiratory Defence." *European Respiratory Journal* 9 (1996):371–76.

Eisen, J. A. *The Tree of Life Blog.* "#badomics words." 2016. phylogenomics.blogspot.com/p/my-writings-on-badomics-words.html.

El-Sayed, Ahmed F. *Fundamentals of Aircraft and Rocket Propulsion.* New York: Springer, 2016.

Engber, Daniel. "The Neurologist Who Hacked His Brain—And Almost Lost His Mind." *Wired,* January 26, 2016. wired.com/2016/01/phil-kennedy-mind-control-computer.

Environmental Protection Agency "Sources of Greenhouse Gas Emissions." 2016. epa.gov/ghgemissions/sources-greenhouse-gas-emissions.

EUROFusion. "JET: Europe's Largest Fusion Device—Funded and Used in Partnership." 2016. euro-fusion.org/jet.

Everett, Daniel L. *Don't Sleep, There Are Snakes: Life and Language in the Amazonian Jungle.* New York: Vintage, 2009.

Fitzpatrick, Michael. "A Long Road for High-Speed Maglev Trains in the U.S." *Fortune,* February 6, 2014. fortune.com/2014/02/06/a-long-road-for-high-speed-maglev-trains-in-the-u-s.

Francis, John. "Diving Impaired: Nitrogen Narcosis." ScubaDiving.com. March 14, 2007. scubadiving.com/training/basic-skills/diving-impaired.

Frederix, M., Mingardon, F., Hu, M., Sun, N., Pray, T., Singh, S., Simmons, B. A., Keasling, J. D., and Mukhopadhyay, A. "Development of an E. coli Strain for One-Pot Biofuel Production from Ionic Liquid Pretreated Cellulose and Switchgrass." *Green Chemistry* 18, no. 15 (2016):4189–97.

Freidberg, Jeffrey P. *Plasma Physics and Fusion Energy.* Cambridge: Cambridge University Press, 2008.

Fusor.net. fusor.net.

Futron Corporation. "Space Transportation Costs: Trends in Price Per Pound to Orbit 1990-2000." Bethesda, Md.: Futron Corporation, 2002.

Garcia, Mark. "Facts and Figures." NASA. 2016. nasa.gov/feature/facts-and-figures.

Gasson, Mark N., Kosta, E., and Bowman, Diana M. *Human ICT Implants: Technical, Legal and Ethical Considerations.* The Hague, The Netherlands: T.M.C. Asser Press, 2012.

Gay, Malcolm. *The Brain Electric: The Dramatic High-Tech Race to Merge Minds and Machines.* New York: Farrar, Straus and Giroux, 2015.

General Fusion. *Rethink Fusion Blog.* generalfusion.com/category/blog.

Gleick, James. "In the Trenches of Science." *New York Times Magazine,* August 16, 1987. nytimes.com/1987/08/16/magazine/in-the-trenches-of-science.html.

Glieder, A., Kubicek, C. P., Mattanovich, D., Wilhschi, B., and Sauer, M. *Synthetic Biology.* New York: Springer, 2015.

Glover, Asha. "NRC's 'All or Nothing' Licensing Process Doesn't Work, Former Commissioner Says." Morning Consult.com, April 29, 2016. morningconsult.com/alert/nrcs-nothing-licensing-process-doesnt-work-former-commissioner-says.

Goodman, Daniel, and Angelova, Kamelia. "TECH STAR: I Want To Punch Anyone Wearing Google Glass in the Face." *BusinessInsider,* May 10, 2013. businessinsider.com/meetup-ceo-scott-heiferman-on-google-glass-2013-5. (Note: The video on this page is no longer working).

Graber, John. "SpriteMods.com's 3D Printer Makes Food Dye Designs in JELLO." 3D Printer World. January 4, 2014. 3dprinterworld.com/article/spritemodscoms-3d-printer-makes-food-dye-designs-jello.

Gramazio, Fabio, and Kohler, Matthias, ed. "Special Issue: Made by Robots: Challenging Architecture at a Larger Scale." *Architectural Design* 84, no. 3 (2014):136.

Grant, Dale. *Wilderness of Mirrors: The Life of Gerald Bull.* Scarborough, Ont.: Prentice Hall, 1991.

Green, Keith Evan. *Architectural Robotics: Ecosystems of Bits, Bytes, and Biology,* Cambridge, MA: MIT Press, 2016.

Greenpeace. "Nuclear Fusion Reactor Project in France: An Expensive and Senseless Nuclear Stupidity." Greenpeace International press release. June 28, 2005. www.greenpeace.org/international/en/press/releases/2005/ITERprojectFrance.

Gribbin, John. *Computing with Quantum Cats: From Colossus to Qubits.* Amherst, N.Y.: Prometheus Books, 2014.

Guger, C., Müller-Putz, G., and Allison, B. *Brain-Computer Interface Research: A State-of-the-Art Summary 4.* New York: Springer, 2016.

Hall, Loura. "3D Printing: Food in Space." NASA. July 28, 2013. nasa.gov/directorates/spacetech/home/feature_3d_food.html.

BIBLIOGRAPHY

Hall, Stephen S. "Daniel Nocera: Maverick Inventor of the Artificial Leaf." Innovators. *National Geographic*, May 19, 2014. news.nationalgeographic.com/news/innovators/2014/05/140519-nocera-chemistry-artificial-leaf-solar-renewable-energy.

Hammond, A., Galizi, R., Kyrou, K., Simoni, A., Siniscalchi, C., Katsanos, D., Gribble, M., Baker, D., Marois, E., Russell, S., et al. "A CRISPR-Cas9 Gene Drive System Targeting Female Reproduction in the Malaria Mosquito Vector Anopheles *gambiae*." *Nature Biotechnology* 34, no. 1 (2016): 78–83.

Hannemann, Christine. *Die Platte Industrialisierter Wohnungsbau in der DDR*. Braunschweig/Wiesbaden: Friedr, Vieweg & Sohn Verlagsgesellschaft mbH, 1996.

Hardesty, Larry. "Ingestible Origami Robot." MIT News. May 12, 2016. news.mit.edu/2016/ingestible-origami-robot-0512.

Harris, A. F., McKemey, A. R., Nimmo, D., Curtis, Z., Black, I., Morgan, S. A., Oviedo, M. N., Lacroix, R., Naish, N., Morrison, N. I., et al. "Successful Suppression of a Field Mosquito Population by Sustained Release of Engineered Male Mosquitoes." *Nature Biotechnology* 30, no. 9 (2012):828–830

Harris, A. F., Nimmo, D., McKemey, A. R., Kelly, N., Scaife, S., Donnelly, C. A., Beech, C., Petrie, W. D., and Alphey, L. "Field Performance of Engineered Male Mosquitoes." *Nature Biotechnology* 29, no. 11 (2011):1034–37.

Hartley, L. P. *The Go-Between*. New York: CA: NYRB Classics, 2002.

Harwood, W. "Experts Applaud SpaceX Rocket Landing, Potential Savings." CBS News, December 22, 2015. cbsnews.com/news/experts-applaud-spacex-landing-cautious-about-outlook.

Hassanien, A. E., and Azar, A. T. *Brain-Computer Interfaces: Current Trends and Applications*. New York: Springer, 2014

Hawkes, E., An, B., Benbernou, N. M., Tanaka, H., Kim, S., Demaine, E. D., Rus, D., and Wood, R. J. "Programmable Matter by Folding." *Proceedings of the National Academy of Sciences* 107, no. 28 (2010):12441–445.

Heaps, Leo. *Operation Morning Light: Inside Story of Cosmos 954 Soviet Spy Satellite*. N.p: Paddington, 1978.

Henderson, D. A., and Preston, Richard. *Smallpox: The Death of a Disease—The Inside Story of Eradicating a Worldwide Killer*. Amherst, N.Y: Prometheus Books, 2009.

Hill, Curtis. *What If We Made Space Travel Practical—Stimulating Our Economy with New Technology*. N.p.: Modern Millennium Press, 2013.

Hoyt, Robert. "WRANGLER: Capture and De-Spin of Asteroids and Space Debris." NASA. May 30, 2014. nasa.gov/content/wrangler-capture-and-de-spin-of-asteroids-and-space-debris.

Hrala, Josh. "This Robot Keeps Trying to Escape a Lab in Russia." Science Alert. June 29, 2016. sciencealert.com/the-same-robot-keeps-trying-to-escape-a-lab-in-russia-even-after-reprogramming.

Hu, Z., Chen, X., Zhao, Y., Tian, T., Jin, G., Shu, Y., Chen, Y., Xu, L., Zen, K., Zhang, C., et al. "Serum MicroRNA Signatures Identified in a Genome Wide Serum MicroRNA Expression Profiling Predict Survival of Non-Small-Cell Lung Cancer." *Journal of Clinical Oncology* 28, no. 10 (2010):1721–26.

Hutchison, C. A., Chuang, R.-Y., Noskov, V. N., Assad-Garcia, N., Deerinck, T. J., Ellisman, M. H., Gill, J., Kannan, K., Karas, B. J., Ma, L., et al. "Design and Synthesis of a Minimal Bacterial Genome." *Science* 351, no. 6253 (2016):aad6253.

iGEM. 2016. igem.org/Main_Page.

Illusio, Inc. "Augmented Reality Goes Beyond Pokemon Go into Plastic Surgery Imaging." Illusio press release. Updated August 5, 2016. www.newswire.com/news/augmented-reality-goes-beyond-pokemon-go-into-plastic-surgery-imaging-13533198.

Innovega Inc. 2015. innovega-inc.com.

Interlandi, Jeneen "The Paradox of Precision Medicine." *Scientific American*, April 1, 2016. scientificamerican.com/article/the-paradox-of-precision-medicine.

International Space Elevator Consortium. "Space Elevator Home." 2016. isec.org.

ITER. "The Way to New Energy." 2016. iter.org.

Jafarpour, F., Biancalani, T., and Goldenfeld, N. "Noise-Induced Mechanism for Biological Homochirality of Early Life Self-Replicators." *Physical Review Letters* 115, no. 15 (2015):158101.

Jain, Kewal K. *Textbook of Personalized Medicine*. New York: Humana Press, 2015.

JAXA. "3-2-2-1 Settlement of Claim between Canada and the Union of Soviet Socialist Republics for Damage Caused by Cosmos 954.'" Japan Aerospace Exploration Agency. Released on April 2, 1981. www.jaxa.jp/library/space_law/chapter_3/3-2-2-1_e.html.

Jella, S. A., and Shannahoff-Khalsa, D. S. "The Effects of Unilateral Forced Nostril Breathing on Cognitive Performance." *International Journal of Neuroscience* 73, no. 1–2 (1993a):61–68.

Jinek, M., Chylinski, K., Fonfara, I., Hauer, M., Doudna, J. A., and Charpentier, E. "A Programmable Dual-RNA–Guided DNA Endonuclease in Adaptive Bacterial Immunity." *Science* 337, no, 6096 (2012):816–21.

Johnson, Aaron. "How Many Solar Panels Do I Need on My House to Become Energy Independent?" Ask an Engineer. MIT School of Engineering. November 19, 2013. engineering.mit.edu/ask/how-many-solar-panels-do-i-need-my-house-become-energy-independent.

BIBLIOGRAPHY

Johnson, L. A., Scholler, J., Ohkuri, T., Kosaka, A., Patel, P. R., McGettigan, S. E., Nace, A. K., Dentchev, T., Thekkat, P., Loew, A., et al. "Rational Development and Characterization of Humanized Anti–EGFR Variant III Chimeric Antigen Receptor T Cells for Glioblastoma." *Science Translational Medicine* 7, no. 275 (2015):275ra22–275ra22.

Josephson, B. "Brian Josephson's home page." Cavendish Laboratory, University of Cambridge. 2016. www.tcm.phy.cam.ac.uk/~bdj10.

Josephson, B. "Brian Josephson on the Memory of Water." Hydrogen2Oxygen. October 6, 2016. hydrogen2oxygen.net/en/brian-josephson-on-the-memory-of-water.

Kaiser, Jocelyn, and Normile, Dennis. "Chinese Paper on Embryo Engineering Splits Scientific Community." *Science.* April 24, 2015. sciencemag.org/news/2015/04/chinese-paper-embryo-engineering-splits-scientific-community.

Kareklas, K., Nettle, D., and Smulders, T. V. "Water-Induced Finger Wrinkles Improve Handling of Wet Objects." *Biology Letters* 9, no. 2 (2013):20120999.

Kaufman, Scott. *Project Plowshare: The Peaceful Use of Nuclear Explosives in Cold War America.* Ithaca, N.Y.: Cornell University Press, 2012.

Kharecha, P. A., and Hansen, J. E. "Prevented Mortality and Greenhouse Gas Emissions from Historical and Projected Nuclear Power." *Environmental Science & Technology* 47, no. 9 (2013):4889–95.

Khoshnevis, Behrokh. "Contour Crafting Simulation Plan for Lunar Settlement Infrastructure Build-Up." NASA Space Technology Mission Directorate. (2013) nasa.gov/directorates/spacetech/niac/khoshnevis_contour_crafting.html.

Kipper, Greg, and Rampolla, Joseph. *Augmented Reality: An Emerging Technologies Guide to AR.* Amsterdam and Boston, Mass.: Syngress, 2012.

Kirsch, Scott. *Proving Grounds: Project Plowshare and the Unrealized Dream of Nuclear Earthmoving.* New Brunswick, N.J.: Rutgers University Press, 2005.

Kolarevic, Branko, and Parlac, Vera. *Building Dynamics: Exploring Architecture of Change.* London and New York: Routledge, 2015.

Kotula, J. W., Kerns, S. J., Shaket, L. A., Siraj, L., Collins, J. J., Way, J. C., and Silver, P. A. "Programmable Bacteria Detect and Record an Environmental Signal in the Mammalian Gut." *Proceedings of the National Academy of Sciences* 111, no. 13 (2014):4838–43.

Kozomara, Ana, and Griffiths-Jones, Sam. "miRBase: Annotating High Confidence MicroRNAs Using Deep Sequencing Data." *Nucleic Acids Research* 42, no. D1 (2014):D68–D73.

Kremeyer, K., Sebastian, K., and Shu, C. "Demonstrating Shock Mitigation and Drag Reduction by Pulsed energy Lines with Multi-domain WENO." Brown University. brown.edu/research/projects/scientific-computing/sites/brown.edu.research.projects.scientific-computing/files/uploads/Demonstrating%20Shock%20Mitigation%20and%20Drag%20Reduced%20by%20Pulsed%20Energy%20Lines.pdf

LaFrance, Adrienne. "Genetically Modified Mosquitoes: What Could Possibly Go Wrong?" *Atlantic,* April 26, 2016. theatlantic.com/technology/archive/2016/04/genetically-modified-mosquitoes-zika/479793.

Lawrence Livermore National Laboratory. "Lasers, Photonics, and Fusion Science: Bringing Star Power to Earth." lasers.llnl.gov/.

LeCroy, C., Masiello, C. A., Rudgers, J. A., Hockaday, W. C., and Silberg, J. J. "Nitrogen, Biochar, and Mycorrhizae: Alteration of the Symbiosis and Oxidation of the Char Surface." *Soil Biology and Biochemistry* 58 (2013):248–54.

Lefaucheur, J.-P., André-Obadia, N., Antal, A., Ayache, S. S., Baeken, C., Benninger, D. H., Cantello, R. M., Cincotta, M., de Carvalho, M., De Ridder, D., et al. "Evidence-Based Guidelines on the Therapeutic Use of Repetitive Transcranial Magnetic Stimulation (rTMS)." *Clinical Neurophysiology* 125, no. 11 (2014):2150–2206.

Levin, Gilbert V. Sweetened Edible Formulations. U.S. Patent Application US 05/838,211, filed September 30, 1977.4262032A. Google Patents. google.com/patents/US4262032.

Lewis, John S. *Asteroid Mining 101: Wealth for the New Space Economy.* Mountain View, Calif.: Deep Space Industries, 2014.

Liang, P., Xu, Y., Zhang, X., Ding, C., Huang, R., Zhang, Z., Lv, J., Xie, X., Chen, Y., Li, Y., et al. "CRISPR/Cas9-Mediated Gene Editing in Human Tripronuclear Zygotes." *Protein and Cell* 6, no. 5 (2015):363–72.

Lipson, Hod, and Kurman, Melba. *Fabricated: The New World of 3D Printing.* Indianapolis, Ind.: Wiley, 2013.

Lockheed Martin. "Compact Fusion." 2016. lockheedmartin.com/us/products/compact-fusion.html.

Lowther, William. *Arms and the Man: Dr. Gerald Bull, Iraq and the Supergun.* Novato, Calif.: Presidio Press, 1992.

Maeda, Junichiro. "Current Research and Development and Approach to Future Automated Construction in Japan," In *Construction Research Congress: Broadening Perspectives,* 1–11. Reston, Va,: American Society of Civil Engineers, 2005. Published online April 26, 2012.

Mahaffey, James A. *Fusion.* New York: Facts on File, 2012a.

MakerBot. "Frostruder MK2." Thingiverse. November 2, 2009. thingiverse.com/thing:1143.

Mali, P., Yang, L., Esvelt, K. M., Aach, J., Guell, M., DiCarlo, J. E., Norville, J. E., and Church, G. M. "RNA-Guided Human Genome Engineering via Cas9." *Science* 339, no. 6121 (2013):823–26.

Malyshev, D. A., Dhami, K., Lavergne, T., Chen, T., Dai, N., Foster, J. M., Corrêa, I. R., and Romesberg, F. E. "A Semi-Synthetic Organism with an Expanded Genetic Alphabet." *Nature* 509, no. 7500. (2014):385–88.

Mankins, John. *The Case for Space Solar Power.* Houston, Tex.: Virginia Edition Publishing, 2014.

Mann Library. "Fast and Affordable: Century of Prefab Housing. Thomas Edison's Concrete House." Cornell University. 2006. exhibits.mannlib.cornell.edu/prefabhousing/prefab.php?content=two_a.

Mannoor, M. S., Jiang, Z., James, T., Kong, Y. L., Malatesta, K. A., Soboyejo, W. O., Verma, N., Gracias, D. H., and McAlpine, M. C. "3D Printed Bionic Ears." *Nano Letters,* 13, no. 6 (2013):2634–39.

Mark Foster Gage Architects. "Robotic Stone Carving." mfga.com/robotic-stone-carving.

Markstedt, K., Mantas, A., Tournier, I., Martínez Ávila, H., Hägg, D., and Gatenholm, P. "3D Bioprinting Human Chondrocytes with Nanocellulose–Alginate Bioink for Cartilage Tissue Engineering Applications." *Biomacromolecules* 16, no. 5 (2015):1489–96.

Mars One. mars-one.com

Mayo Clinic Staff. "Transcranial Magnetic Stimulation—Overview." Mayo Clinic. 2015. mayoclinic.org/tests-procedures/transcranial-magnetic-stimulation/home/ovc-20163795.

McCracken, Garry, and Stott, Peter. *Fusion: The Energy of the Universe.* Cambridge, Mass.: Academic Press, 2012.

McGee, Ellen M., and Maguire, Gerald Q. "Becoming Borg to Become Immortal: Regulating Brain Implant Technologies." *Cambridge Quarterly of Healthcare Ethics* 16, no. 3 (2007):291–302.

McNab, I. R. "Launch to Space with an Electromagnetic Railgun." *IEEE Transactions on Magnetics* 39, no. 1 (2003): 295–304.

Menezes, A. A., Cumbers, J., Hogan, J. A., and Arkin, A. P. "Towards Synthetic Biological Approaches to Resource Utilization on Space Missions." *Journal of the Royal Society Interface* 12, no. 1-2 (2015):20140715.

Miller, Jordan S. "The Billion Cell Construct: Will Three-Dimensional Printing Get Us There?" *PLoS Biology* 12, no. 6 (2014):e1001882.

MIT Technology Review. (November 2012): 115, 108. technologyreview.com/magazine/2012/11/.

Moan, Charles E., and Heath, Robert G. "Septal stimulation for the Initiation of Heterosexual Behavior in a Homosexual Male." *Journal of Behavior Therapy and Experimental Psychiatry* 3, no. 1 (1972):23–30.

Mohan, Pavithra. "App Used 23andMe's DNA Database to Block People From Sites Based on Race and Gender." *Fast Company.* July 23, 2015. fastcompany.com/3048980/fast-feed/app-used-23andmes-dna-database-to-block-people-from-sites-based-on-race-and-gender.

Mohiuddin, M. M., Singh, A. K., Corcoran, P. C., Hoyt, R. F., Thomas III, M. L., Ayares, D., and Horvath, K. A. "Genetically Engineered Pigs and Target-Specific Immunomodulation Provide Significant Graft Survival and Hope for Clinical Cardiac Xenotransplantation." *Journal of Thoracic and Cardiovascular Surgery* 148, no. 3 (2011).1106–14.

Molloy, Mark. "Hiroshima Anger Over Pokémon at Atom Bomb Memorial Park." *Telegraph,* July 28, 2016. telegraph.co.uk/technology/2016/07/28/hiroshima-anger-over-pokemon-at-atom-bomb-memorial-park.

Moniz, E. J. "U.S. Participation in the ITER Project." Washington, D.C.: United States Department of Energy, 2016. science.energy.gov/~/media/fes/pdf/DOE_US_Participation_in_the_ITER_Project_May_2016_Final.pdf.

Moravec, H. *Mind Children: The Future of Robot and Human Intelligence.* Cambridge, Mass.: Harvard University Press, 1990.

Moser, M. B., and Moser, E. I. *The Future of the Brain: Essays by the World's Leading Neuroscientists.* Princeton, N.J.: Princeton University Press, 2014a.

———. *The Future of the Brain: Essays by the World's Leading Neuroscientists.* Princeton, N.J.: Princeton University Press, 2014b.

Moskvitch, K. "Programmable Matter: Shape-Shifting Microbots Get It Together." *Engineering and Technology Magazine* 10, no. 5 (2015). eandt.theiet.org/content/articles/2015/05/programmable-matter-shape-shifting-microbots-get-it-together.

Mourachkine, Andrei. *Room-Temperature Superconductivity.* Cambridge, U.K.: Cambridge International Science Publishing, 2004.

Mukherjee, Siddhartha. *The Emperor of All Maladies: A Biography of Cancer.* New York: Scribner, 2011.

Muller, Richard A. *Energy for Future Presidents: The Science Behind the Headlines.* New York: W. W. Norton, 2013.

Mullin, Rick. "Cost to Develop New Pharmaceutical Drug Now Exceeds $2.5B." *Scientific American,* November 24, 2014. www.scientificamerican.com/article/cost-to-develop-new-pharmaceutical-drug-now-exceeds-2-5b/.

Murphy, Sean V., and Atala, Anthony. "3D Bioprinting of Tissues and Organs." *Nat Biotech* 32, no. 8 (2014):773–85.

Naboni, Roberto, and Paoletti, Ingrid. *Advanced Customization in Architectural Design and Construction.* New York: Springer, 2014.

Naclerio, R. M., Bachert, C., and Baraniuk, J. N. "Pathophysiology of Nasal Congestion." *International Journal of General Medicine* 3, (2010):7–57.

NASA. "A Natural Way To Stay Sweet." NASA Spinoff. 2004. spinoff.nasa.gov/Spinoff2004/ch_4.html

———. "Welcome to the Dawn Mission." NASA Jet Propulsion Laboratory. N.d. dawn.jpl.nasa.gov/mission.

National Academies of Sciences, Engineering, and Medicine. *Gene Drives on the Horizon: Advancing Science, Navigating Uncertainty, and Aligning Research with Public Values.* Washington, D.C.: National Academies Press, 2016.

National Cancer Institute. "SEER Stat Fact Sheets: Cancer of the Lung and Bronchus." 2013. seer.cancer.gov/statfacts/html/lungb.html.

National Institutes of Health. "Precision Medicine Initiative." 2015. www.nih.gov/precision-medicine-initiative-cohort-program.

NeuroPace, Inc. "The RNS® System for Drug-Resistant Epilepsy." NeuroPace, Inc. 2016. neuropace.com.

New Age Robotics. "Milling & Sculpting." 2016. robotics.ca/wp/portfolio/milling-sculpting.

Newman, A. M., Bratman, S. V., To, J., Wynne, J. F., Eclov, N. C. W., Modlin, L. A., Liu, C. L., Neal, J. W., Wakelee, H. A., Merritt, R. E., et al. "An Ultrasensitive Method for Quantitating Circulating Tumor DNA with Broad Patient Coverage." *Nature Medicine* 20, no. 5 (2014):548–54.

Northrop, Robert B., and Connor, Anne N. *Ecological Sustainability: Understanding Complex Issues.* Boca Raton, FL: CRC Press, 2013.

Obed, A., Stern, S., Jarrad, A., and Lorf, T. "Six Month Abstinence Rule for Liver Transplantation in Severe Alcoholic Liver Disease Patients." *World Journal of Gastroenterology* 21, no. 14 (2015):4423–26.

offensive-computing (username). "Genetic Access Control." 2015. https://github.com/offapi/rbac-23andme-oauth2.

Orlando, L., Ginolhac, A., Zhang, G., Froese, D., Albrechtsen, A., Stiller, M., Schubert, M., Cappellini, E., Petersen, B., Moltke, I., et al. "Recalibrating Equus Evolution Using the Genome Sequence of an Early Middle Pleistocene Horse." *Nature* 499, no. 7457 (2013):74–78.

Open Humans. openhumans.org.

Organovo. "Bioprinting Functional Human Tissue." 2016. organovo.com.

Owen, David. *The Conundrum.* New York: Riverhead Books, 2012.

Paddon, C. J., Westfall, P. J., Pitera, D. J., Benjamin, K., Fisher, K., McPhee, D., Leavell, M. D., Tai, A., Main, A., Eng, D., et al. "High-Level Semi-Synthetic Production of the Potent Antimalarial Artemisinin." *Nature* 496, no. 7446 (2013):528–32.

Pais-Vieira, M., Chiuffa, G., Lebedev, M., Yadav, A., and Nicolelis, M. A. L. "Building an Organic Computing Device with Multiple Interconnected Brains." *Scientific Reports* 5 (2015):11869.

Pelt, Michel van. *Rocketing into the Future: The History and Technology of Rocket Planes.* New York: Springer, 2012.

———. *Space Tethers and Space Elevators.* New York: Copernicus, 2009.

Peplow, Mark. "Synthetic Biology's First Malaria Drug Meets Market Resistance: Nature News & Comment." *Nature* 530, no. 7591 (2016):389–90.

Perez, Sarah. "Recognizr: Facial Recognition Coming to Android Phones." ReadWrite. February 24, 2010. readwrite.com/2010/02/24/recognizr_facial_recognition_coming_to_android_phones.

Personal Genome Project. "Sharing Personal Genomes." Personal Genome Project: Harvard Medical School. personalgenomes.org.

Phillips, Tony. "The Tunguska Impact—100 Years Later." NASA Science. 2008. science.nasa.gov/science-news/science-at-nasa/2008/30jun_tunguska.

Phipps, C., Birkan, M., Bohn, W., Eckel, H.-A., Horisawa, H., Lippert, T., Michaelis, M., Rezunkov, Y., Sasoh, A., Schall, W., et al. "Review: Laser-Ablation Propulsion." *Journal of Propulsion and Power* 26, no. 4 (2010):609–37.

Pino, R. E., Kott, A., Shevenell, M., ed. *Cybersecurity Systems for Human Cognition Augmentation.* New York: Springer, 2014.

Piore, Adam. "To Study the Brain, a Doctor Puts Himself Under the Knife." *MIT Technology Review.* November 9, 2015. technologyreview.com/s/543246/to-study-the-brain-a-doctor-puts-himself-under-the-knife.

Pleistocene Park. "Pleistocene Park: Restoration of the Mammoth Steppe Ecosystem." 2016. pleistocenepark.ru/en.

Poland, Gregory A. "Vaccines Against Lyme Disease: What Happened and What Lessons Can We Learn?" *Clinical Infectious Diseases* 52, supp. 3 (2011):s253–s258.

Polka, Jessica K., and Silver, Pamela A. "A Tunable Protein Piston That Breaks Membranes to Release Encapsulated Cargo." *ACS Synthetic Biology* 5, no. 4 (2016):303–11.

Post, Hannah. "Reusability: The Key to Making Human Life Multi-Planetary." SpaceX. June 10, 2015. spacex.com/news/2013/03/31/reusability-key-making-human-life-multi-planetary.

Powell, J., Maise, G., and Pellegrino, C. *StarTram: The New Race to Space* (N. p.: CreateSpace Independent Publishing Platform, 2013.

Rabinowits, G., Gerçel-Taylor, C., Day, J. M., Taylor, D. D., and Kloecker, G. H. "Exosomal MicroRNA: A Diagnostic Marker for Lung Cancer." *Clinical Lung Cancer* 10, no. 1 (2009):42–46.

Reaction Engines Limited. reactionengines.co.uk.

Reardon, Sara. "New Life for Pig-to-Human Transplants." *Nature* 527, no. 7577 (2015):152–54.

Reece, Andrew G., and Danforth, Christopher M. "Instagram Photos Reveal Predictive Markers of Depression." arXiv:1608.03282 [physics] (2016):34.

Reece, A. G., Reagan, A. J., Lix, K. L. M., Dodds, P. S., Danforth, C. M., and Langer, E. J. "Forecasting the Onset and Course of Mental Illness with Twitter Data." arXiv:1608.07740 [physics] (2016): 23.

Reiber, C., Shattuck, E. C., Fiore, S., Alperin, P., Davis, V., and Rowe, J. "Change in Human Social Behavior in Response to a Common Vaccine." *Annals of Epidemiology* 20, no 10 (2010):729–33.

Reid, G., Kirschner, M. B., and van Zandwijk, N. "Circulating microRNAs: Association with Disease and Potential Use As Biomarkers." *Critical Reviews in Oncology/Hematology* 80, no. 2 (2011):193–208.

Reiss, Louise Z. "Strontium-90 Absorption by Deciduous Teeth." *Science* 134, no 3491 (1961):1669–73.

Riaz, Muhammad U., and Javaid, Zain. *Programmable Matter: World with Controllable Matter.* Saarbrücken, Germany: Lambert Academic Publishing, 2012.

Richter, B., and Neises, G. "'Human' Insulin Versus Animal Insulin in People with Diabetes Mellitus." Cochrane Database of Systematic Reviews (2002):CD003816.

Ringeisen, B., Spargo, B. J., and Wu, Peter K. *Cell and Organ Printing.* Dordrecht, Germany: Springer, 2010.

Ringo, Allegra. "Understanding Deafness: Not Everyone Wants to Be 'Fixed.'" *Atlantic.* August 9, 2013. theatlantic.com/ health/archive/2013/08/understanding-deafness-not-everyone-wants-to-be-fixed/278527.

Robinette, Paul. "Developing Robots That Impact Human Robot Trust In Emergency Evacuations." PhD thesis. Georgia Institute of Technology, 2015.

Romanishin, J. W., Gilpin, K., and Rus, D. "M-Blocks: Momentum-Driven, Magnetic Modular Robots," 4288–95. *IEEE/RSJ International Conference on Intelligent Robots and Systems,* Piscataway, N.J.: IEEE Publishing, 2013.

Rose, David. *Enchanted Objects: Innovation, Design, and the Future of Technology.* New York: Scribner, 2015.

Roth, Alvin E. *Who Gets What—and Why: The New Economics of Matchmaking and Market Design.* Boston: Eamon Dolan/ Mariner Books, 2016.

Rubenstein, M. "Emissions from the Cement Industry." *State of the Planet.* Earth Institute. Columbia University. May 9, 2012. blogs.ei.columbia.edu/2012/05/09/emissions-from-the-cement-industry.

Rubenstein, M., Cornejo, A., and Nagpal, R. "Programmable Self-Assembly in a Thousand-Robot Swarm." *Science* 345, no, 6198 (2014):795–99.

Sandia National Laboratories. "Sandia Magnetized Fusion Technique Produces Significant Results." September 22, 2014. share.sandia.gov/news/resources/news_releases/mag_fusion/#.V8GkOpMrJE5.

———. "Z Pulsed Power Facility." Z Research: Energy. 2015. www.sandia.gov/z-machine/research/energy.html.

Schafer, G., Green, K., Walker, I., King Fullerton, S., and Lewis, E. "An Interactive, Cyber-Physical Read-Aloud Environment: Results and Lessons from an Evaluation Activity with Children and Their Teachers," 865–74. In *Proceedings of the 2014 Conference on Designing Interactive Systems.* New York: ACM, 2014.

Schulz, A., Sung, C., Spielberg, A., Zhao, W., Cheng, Y., Mehta, A., Grinspun, E., Rus, D., and Matusik, W. "Interactive Robogami: Data Driven Design for 3D Print and Fold Robots with Ground Locomotion." 1:1. In *SIGGRAPH 2015: Studio.* New York: ACM, 2015.

Schwenk, Kurt. "Why Snakes Have Forked Tongues." *Science* 263, no. 1573 (1994):1573–77.

Seife, Charles. *Sun in a Bottle: The Strange History of Fusion and the Science of Wishful Thinking.* New York: Viking, 2008.

Seiler, Friedrich, and Igra, Ozer. *Hypervelocity Launchers.* New York: Springer, 2016.

Selectbio. "Caddie Wang's Biography." 3D-Printing in Life Sciences. Selectbio Sciences. 2015. selectbiosciences.com/ conferences/biographies.aspx?speaker=1340332&conf=PRINT2015.

Self-Assembly Lab. selfassemblylab.net/index.php.

Sepramaniam, S., Tan, J.-R., Tan, K.-S., DeSilva, D. A., Tavintharan, S., Woon, F.-P., Wang, C.-W., Yong, F. L., Karolina, D.-S., Kaur, P., et al. "Circulating MicroRNAs as Biomarkers of Acute Stroke." *International Journal of Molecular Sciences* 15, no. 1 (2014):1418–32.

Serafini, G., Pompili, M., Belvederi Murri, M., Respino, M., Ghio, L., Girardi, P., Fitzgerald, P. B., and Amore, M. "The Effects of Repetitive Transcranial Magnetic Stimulation on Cognitive Performance in Treatment-Resistant Depression. A Systematic Review." *Neuropsychobiology* 71, no. 3 (2015):125–39.

Sercel, Joel. "APIS (Asteroid Provided In-Situ Supplies): 100MT Of Water from a Single Falcon 9." NASA. May 7, 2015. nasa.gov/feature/apis-asteroid-provided-in-situ-supplies-100mt-of-water-from-a-single-falcon-9.

Servick, Kelly. "Scientists Reveal Proposal to Build Human Genome from Scratch." *Science,* June 2, 2016, sciencemag.org/ news/2016/06/scientists-reveal-proposal-build-human-genome-scratch.

Shapiro, Beth. *How to Clone a Mammoth: The Science of De-Extinction.* Princeton, N.J.: Princeton University Press, 2015.

Shine, Richard, and Wiens, John J. "The Ecological Impact of Invasive Cane Toads (*bufo marinus*) in Australia." *Quarterly Review of Biology* 85, no. 3 (2010):253–91.

Shreeve, James. *The Genome War: How Craig Venter Tried to Capture the Code of Life and Save the World.* New York: Knopf, 2004.

Silverstein, Ken. "How the Chips Fell." *Mother Jones* 22, (1997):13–14.

Simberg, Rand E., and Lu, Ed. *Safe Is Not an Option* New York: Interglobal Media LLC, 2013.

Small, E. M., and Olson, E. N. "Pervasive Roles of microRNAs in Cardiovascular Biology." *Nature* 469, no. 7330. (2011):336–42.

Smith, Dan. "DARPA's 'Programmable Matter' Project Creating Shape-Shifting Materials." *Popular Science*, June 8, 2009. popsci.com/military-aviation-amp-space/article/2009-06/mightily-morphing-powerful-range-objects.

Snir, A., Nadel, D., Groman-Yaroslavski, I., Melamed, Y., Sternberg, M., Bar-Yosef, O., and Weiss, E. "The Origin of Cultivation and Proto-Weeds, Long Before Neolithic Farming." *PLOS ONE* 10, no. 7 (2015):e0131422.

Snyder, Michael. *Genomics and Personalized Medicine: What Everyone Needs to Know.* Oxford and New York: Oxford University Press, 2016.

Somlai-Fischer, A., Hasegawa, A., Jasinowicz, B., Sjölén, T., and Hague, U. "Reconfigurable House." 2008. http://house. propositions.org.uk.

Spaceflight101. "Falcon 9 v1.1 & F9R—Rockets." 2016a. spaceflight101.com/spacerockets/falcon-9-v1-1-f9r/.

———. "Soyuz FG—Rockets." 2016b. spaceflight101.com/spacerockets/soyuz-fg.

Spröwitz, A., Moeckel, R., Vespignani, M., Bonardi, S., and Ijspeert, A J. "Roombots: A Hardware Perspective on 3D Self-Reconfiguration and Locomotion with a Homogeneous Modular Robot." *Robotics and Autonomous Systems* 62, no. 7. (2014):1016–33.

Stull, Deborah. "Better Mouse Memory Comes at a Price." *Scientist,* April 2, 2001. the-scientist.com/?articles.view/ articleNo/13302/title/Better-Mouse-Memory-Comes-at-a-Price.

Suthana, Nanthia, and Fried, Itzhak. "Deep Brain Stimulation for Enhancement of Learning and Memory." *NeuroImage* 85, part 3, (2014):996–1002.

Swan, P., Raiit, D., Swan, C., Penny, R., and Knapman, J. *Space Elevators: An Assessment of the Technological Feasibility and the Way Forward.* Paris and Virginia: Science Deck Books, 2013.

Syrian Refugees. "The Syrian Refugee Crisis and Its Repercussions for the EU." 2016. syrianrefugees.eu.

Talbot, David. "A Prosthetic Hand That Sends Feelings to Its Wearer." *MIT Technology Review.* December 5, 2013. technologyreview.com/s/522086/an-artificial-hand-with-real-feelings.

Tan, D. W., Schiefer, M. A., Keith, M. W., Anderson, J. R., Tyler, J., and Tyler, D. J. "A Neural Interface Provides Long-Term Stable Natural Touch Perception." *Science Translational Medicine* 6, no. 257 (2014):257ra138.

Tang, Y.-P., Shimizu, E., Dube, G. R., Rampon, C., Kerchner, G. A., Zhuo, M., Liu, G., and Tsien, J. Z. "Genetic Enhancement of Learning and Memory in Mice." *Nature* 401, no. 6748 (1999):63–69.

Throw Trucks With Your Mind! throwtrucks.com.

Tidball, R., Bluestein, J., Rodriguez, N., and Knoke, S. "Cost and Performance Assumptions for Modeling Electricity Generatin Technologies." Fairfax, Va.: National Renewable Energy Laboratory, 2010. nrel.gov/docs/fy11osti/48595.pdf.

Tinkham, Michael. *Introduction to Superconductivity: Second Edition.* Mineola, N.Y.: Dover Publications, 2004.

Tomich, Jeffrey. "Decades Later, Baby Tooth Survey Legacy Lives On." *St. Louis Post-Dispatch.* August 1, 2013. stltoday. com/lifestyles/health-med-fit/health/decades-later-baby-tooth-survey-legacy-lives-on/article_c5ad9492-fd75-5aed-897f-850fbdba24ee.html.

Torella, J. P., Gagliardi, C. J., Chen, J. S., Bediako, D. K., Colón, B., Way, J. C., Silver, P. A., and Nocera, D. G. "Efficient Solar-to-Fuels Production from a Hybrid Microbial–Water-Splitting Catalyst System." *Proceedings of the National Academy of Sciences* 112, no. 8 (2015):2337–42.

Trang, P. T. K., Berg, M., Viet, P. H., Mui, N. V., and van der Meer, J. R. "Bacterial Bioassay for Rapid and Accurate Analysis of Arsenic in Highly Variable Groundwater Samples." *Environmental Science & Technology.* 39, no. 19 (2005):7625–30.

Treisman, M. "Motion Sickness: An Evolutionary Hypothesis." *Science* 197, no. 4302 (1977):493–95.

UN Habitat. "World Habitat Day: Voices from Slums—Background Paper." United Nations Human Settlements Programme. 2014. unhabitat.org/wp-content/uploads/2014/07/WHD-2014-Background-Paper.pdf.

United Network for Organ Sharing. "Data." 2015. unos.org/data.

U.S. Congress. House. (2008). *Genetic Information Nondiscrimination Act of 2008.* H.R. 493. 110th Cong. *Congressional Record* 154, no. 71, daily ed. (May 1, 2008): H2961–H2980. www.congress.gov/bill/110th-congress/house-bill/493.

———. (2015). *U.S. Commercial Space Launch Competitiveness Act.* H.R. 2262.114th Cong., 1st sess. *Congressional Record* 161, no. 78, daily ed. (May 20, 2015): H3403–H3410. www.congress.gov/bill/114th-congress/house-bill/2262.

Van Nimmen, Jane, Bruno, Leonard C., and Rosholt, Robert L. *NASA Historical Data Book, 1958–1968. Vol I: NASA Resources.* Washington, D.C.: NASA, 1976. history.nasa.gov/SP-4012v1.pdf.

Vasudevan, T. M., van Rij, A. M., Nukada, H., and Taylor, P. K. "Skin Wrinkling for the Assessment of Sympathetic Function in the Limbs." *Australian and New Zealand Journal of Surgery* 70, no. 1 (2000):57–59.

BIBLIOGRAPHY

Venter, J. C. *Life at the Speed of Light: From the Double Helix to the Dawn of Digital Life*. New York: Penguin Books, 2014b.

————. *What—Me Worry?*, 200–07. In *What Should We Be Worried About?: Real Scenarios That Keep Scientists Up at Night*, edited by J. Brockman. New York: Harper Perennial, 2014a.

Vrije Universiteit Science. "Robot Baby Project by Prof.dr. A.E. Eiben on evolving robots / The Evolution of Things." May 26, 2016. youtube.com/watch?v=BfcVSb-Q8ns.

Wang, Brian. "$250,000 Slingatron Kickstarter." NextBigFuture. July 29, 2013. nextbigfuture.com/2013/07/250000-slingatron-kickstarter.html.

Wei, F., Wang, G.-D., Kerchner, G. A., Kim, S. J., Xu, H.-M., Chen, Z.-F., and Zhuo, M. "Genetic Enhancement of Inflammatory Pain by Forebrain NR2B Overexpression." *Natural Neuroscience* 4, no 2 (2001):164–69.

Werfel, Justin. "Building Structures with Robot Swarms." O'Reilly.com. 2016. oreilly.com/ideas/building-structures-with-robot-swarms.

Werfel, J., Petersen, K., and Nagpal, R. "Designing Collective Behavior in a Termite-Inspired Robot Construction Team." *Science* 343, no. 6172 (2014):754–58.

White, D. E., Bartley, J., and Nates, R. J. "Model Demonstrates Functional Purpose of the Nasal Cycle." *BioMedical Engineering OnLine* 14:38 (2015).

Whiting, P., Al, M., Burgers, L., Westwood, M., Ryder, S., Hoogendoorn, M., Armstrong, N., Allen, A., Severens, H., Kleijnen, J., et al. "Ivacaftor for the Treatment of Patients with Cystic Fibrosis and the G551D Mutation: A Systematic Review and Cost-Effectiveness Analysis." *Health Technology Assessment* 18, no. 18 (2014):1–130.

Wikipedia. "Nitrogen Narcosis." 2016. en.wikipedia.org/w/index.php?title=Nitrogen_narcosis&oldid=735322553.

Wittmann, J., and Jäck, H.-M. "Serum microRNAs as Powerful Cancer Biomarkers." *Biochimica et Biophysica Acta (BBA)— Reviews on Cancer* 1806, no. 2 (2010):200–207.

Wolpaw, Jonathan, and Wolpaw, Elizabeth Winter. *Brain-Computer Interfaces: Principles and Practice*. Oxford and New York: Oxford University Press, 2012.

World Health Organization. "Global Insecticide Resistance Database." 2014. who.int/malaria/areas/vector_control/insecticide_resistance_database/en.

————. "10 Facts On Malaria." 2015. who.int/features/factfiles/malaria/en.

World Nuclear Association. "Peaceful Nuclear Explosions." 2010. world-nuclear.org/information-library/non-power-nuclear-applications/industry/peaceful-nuclear-explosions.aspx.

Wrangham, Richard. *Catching Fire: How Cooking Made Us Human*. New York: Basic Books, 2010.

Yang, L., Güell, M., Niu, D., George, H., Lesha, E., Grishin, D., Aach, J., Shrock, E., Xu, W., Poci, J., et al. "Genome-Wide Inactivation of Porcine Endogenous Retroviruses (PERVs)." *Science* 350, no 6264. (2015):1101–1104.

Yim, M., White, P., Park, M., and Sastra, J. (2009). *Modular Self-Reconfigurable Robots*, 5618–31. In *Encyclopedia of Complexity and Systems Science*, edited by Robert A. Meyers. New York: Springer, 2009.

Zewe, Adam. "In Automaton We Trust." 2016. Harvard Paulson School of Engineering and Applied Sciences. seas.harvard.edu/news/2016/05/in-automaton-we-trust.

Zhu, L., Wang, J., and Ding, F. "The Great Reduction of a Carbon Nanotube's Mechanical Performance by a Few Topological Defects." *ACS Nano* 10, no. 6 (2016): 6410–15.

Index

Aaronson, Scott, 330
acceleration, 23–24, 24n, 25
actuators, 105
Adams, James, 49
Africa, 199
afterburner, 21
Airbus A380, 55
Air Force, U.S., 18
airlines, 122
airplanes, 19–20, 29, 122
air resistance, 25–26, 28–29
Alaska, 98
albinism, 196n
alcoholism, 258n
Aldrin, Buzz, 6n
Alexandrians, ancient, 6
algae, 209
Alphabet, 254
Alzheimer's disease, 237, 247, 307
Amazon, 111, 180, 259
ambient markers, 171
amino acids, 193–94, 221, 332
amphetamines, 304, 308
Amyris, Inc., 199
anesthesia, 229
Animated Work Environment, 109
ankles, 323
Antarctica, 53, 327
antigens, 242
Antigua, 48
APIS (Asteroid Provided In-Situ
    Supplies), 63
Apollo 11, 7, 55, 59–60
Apollo space program, 59n, 92
Apple, 272
Architectural Design, 138
architecture, 137–38
"Are Talking Heads Blowing Hot Air"
    (2011 study), 1
Arizona, 320–21
*Arms and the Man: Dr. Gerald Bull, Iraq
    and the Supergun* (Lowther), 50
Army, U.S., 47, 313n
arsenic, 211
art, 183
artemisinin, 198–200

artificial intelligence, 136, 139–40
artificial organs, *see* bioprinting
artspeak, 138
Artsutanov, Yuri, 35
Asian elephants, 223
Asians, 196n
asteroid mining, 52–69, 320n
    benefits of, 68–69
    environmental degradation in,
        66–67
    finances of, 54–56
    law and order in, 65–66
    problems facing, 58–65
    rights to, 63 64
    safety and, 67
asteroid-moving technology, 67
asteroids:
    escape velocity of, 55
    landing on, 62–63
    net capture of, 63
    rubble pile, 62
    types of, 53–54
atmosphere, density of, 25, 29
atomic bombs, 79, 96, 98
atomic gardening, 191–92
ATP, 286
augmented reality (AR), 8n
    accuracy required for, 171
    audio in, 174
    benefits of, 183–86
    concerns about, 180–83
    hacking of, 183
    hardware for, 168
    location detection for, 171–74
    markers in, 169–71
    motion sickness in, 168
    possible uses for, 164–66, 168,
        177–79, 183–86
    reference images in, 172–73
    registration in, 166–68, 172–73
    smell in, 174–75
    vs. virtual reality, 165
    where are we now in, 175–79
Augmented Reality Lab, 173, 177–78
Auschwitz, 183
Australia, 219

automotive industry, 136, 137
autonomous cars, 174

baby teeth, 99
bacteria, 203–5, 206, 210, 218
    in bioprinted organs, 273
    environmental monitoring by,
        211–12
    immune system of, 212–14
    synthetic, 220–21
Bad Astronomy (blog), 36
Barbados, 47
Bartlett School of Graduate Studies,
    141
Baseline Study, 254
bases, 192–93
Bauby, Jean-Dominique, 316
B cell, 242
behavior patterns, 231
Belize, 315, 316
Belleau Wood, Battle of, 178
Berger, Theodore, 308
beryllium, 92
Billinghurst, Mark, 176
bio-ink, 263–66
    components of, 266–67
biomarkers, 230–31, 247
bioprinting, 144, 206, 257–81
    benefits of, 274–75
    concerns about, 272–74
    state of the art in, 268–92
    sugar sintering method in, 270
    two techniques for, 263–66
Biostatistics Research and Consulting
    Center, 235
bioterrorism, 217
birds, 225
Blenner, Mark, 160
blind people, 310
bloodletting, 229
blood type, 195–96
blood vessels, 262, 268
    bioprinting of, 269–71
Bloomberg View, 154
Boeing, 179

351

Bolonkin, Alexander, 62*n*
bomb threats, 130
Booth, Serena, 129–30
Botella, Cristina, 179
Bovine Elite, LLC, 197*n*
brain:
 drugs for modifying, 308
 electric signals in, 285–86
 invasive reading of, 294–95
 metabolic signals in, 286–87
 noninvasive electromagnetic reading
  of, 287–90
 noninvasive metabolic reading of,
  290–93
 optimal conditions for learning in,
  304–5
 reading of, 285–99
 superinvasive reading of, 295–99
 upgrading of, 299–305
 writing to the, 306–8
brain-computer interfaces, 282–317
 benefits of, 311–14
 brain reading and, 285–99
 concerns about, 308–10
 games for, 312
brain-to-brain connection, 312–13
brain tumors, 242–43
Brassica oleracea, 190
breast cancer, 240
breast exams, 238
breeding, 191
brewer's yeast, 199
Brexit, 22*n*
bricklaying, 139–42, 154
Brown University, 28
Brunner, Daniel, 91
Brussels, 50
Bucket of Stuff, 116–20, 124–25, 126,
 128
Bull, Gerald, 45–50
*Bull's Eye: The Life and Times of
 Supergun Inventor Gerald Bull*
 (Adams), 49
Bureau of Labor Statistics, 153, 155
Burj Khalifa, 25*n*
Business Insider, 175
Butcher, Jonathan, 269

calcium, 99
California, University of:
 at Berkeley, 199, 212
 at Davis, 234*n*, 328
 at Santa Barbara, 176
 at Santa Cruz, 222
Canada, 45–47, 48, 58, 321*n*
Canadian Space Society, 53
cancer, 3, 206, 231, 234
 continuing mutation of, 241
 diagnosis of, 238–41
 monitoring of, 243–44
 treatment of, 241–43
cane toads, 219
Canterbury, University of, 176

capillaries, 262, 271
Caplan, Bryan, 56, 154
caraway seeds, 334–35
carbon, 52, 94, 211
carbon dioxide, 208–9, 210
carbon fiber, 143
carbon nanotube, 35–36
cardiac hypertrophy, 246–47
cars, 15, 24*n*
cartilage, 271–72
Case for Space Solar Power, The
 (Mankins), 320
Case Western Reserve University, 151*n*
Cas (protein), 213
cat bricks, 111
CD19 (molecule), 242
Cell and Organ Printing (Ringeisen),
 259
cells, 192–93, 208, 260
 bioprinting and, 264–66
 mutant, 238
cellulose, 210
Center for Smell and Taste, 334
Centers for Disease Control, 217*n*
Ceres, 60
Chagan, Lake, 100
Charpentier, Emmanuelle, 212
ChemBot, 124
chemical loop, 205
chemotherapy, 241, 247
Chicken McNuggets, 193*n*
children, 110–11
children's birthday parties, 178
China, 146, 219, 258
Chinese sweet wormwood plants,
 198–99
chirality, 332–33
Church, George, 203, 214, 220, 223*n*,
 252, 332, 335
CIA (Central Intelligence Agency), 48,
 50
cilia, 187–88
Clemson University, 160
climate change, 41, 94
clothing, 154
cloud cover, 41
CNSA (China National Space
 Administration), 65
coal, 73
"Cobotics," 141*n*
cochlear implants, 306–7, 310
cognitive abilities, 304–5
cold fusion, 5
Cold War, 38
Collins, Francis, 214
Colorado, University of, 176
Comcast, 262
Comic-Con, 78*n*
communications satellites, 34
Complete Anatomy Lab, 185
computerized manufacturing, 137
computers, 2, 101, 139
 brain as, 283–84

prosthetics and, 322–23
 quantum, 328–30
 see also brain-computer interfaces
concrete, 145, 155
Congress, U.S., 18, 64, 250, 274
construction, robotic, *see* robotic
 construction
construction industry, 153–55
Construction Robotics, 141
construction workers, robots as,
 139–44
contact lens, 176
Contour Crafting, 145, 146, 149, 156,
 158
copper, 325
copper wire, 4, 5
Cornell University, 150, 162
cosmetic surgery, 185, 303
Cosmos 954, 58
Coulomb barrier, 77
cows, 210
CPS, 171
Craig, Alan, 182–83, 184
CRISPR-Cas9, 207, 212–14, 219,
 236–37
Crohn's disease, 247
ctDNA (circulating tumor DNA), 240,
 244
C-type (carbonaceous) asteroids, 53
cyborg ear, 271–72
cystic fibrosis, 236–37, 248

D'Andrea, Raffaello, 152
Danforth, Christopher, 247
Daniels, Karen, 63
DAQRI, 179
DARPA (Defense Advanced Research
 Projects Agency), 124
Darth Vader (char.), 324
data encryption, 329
Dawn mission (NASA), 60
deaf people, 310
deep brain stimulation, 299–302, 304,
 309
Deep Space Industries, 53
de-extinction, 221–25
Defense Department, U.S., 47
Delp, Michael, 59*n*
Demaine, Erik, 102, 107–8, 118, 122,
 128
dementia, 307
Dempsey, Gaia, 179
depression, 245, 247, 250, 301, 302
depth perception, via smell, 187
Derleth, Jason, 25–27, 35–36, 40
designer babies, 219
deuterium, 73–74, 77, 83
deuterium gas, 81–82
Deutsch, David, 330
diabetes, 245
diminished reality, 181–82
dinosaurs, 225
disease, 198–203, 217

Disney, Walt, 97
Diving Bell and the Butterfly, The (Bauby), 316
d-limonene, 210
DNA, 191, 192–98, 201–2, 204, 205, 213–14, 217, 221, 222, 234, 236, 239, 332
   of mammoths, 222–23
   as memory storage, 220
Doctor Who (TV show), 82
dogs, 187
Domburg, Jeroen, 161
Dong, Suyang, 177
Doudna, Jennifer, 212
Dowling, Jonathan, 330*n*
drones, 152–53
drugs, 269
drug trials, 254–55, 268–69
Duff, David, 116

ears, 186
Earth, 16, 25, 31, 32, 33, 34, 37, 38, 39, 41, 42, 43, 49, 52, 53, 55, 56, 57, 59, 60, 67, 68, 69, 159, 169, 319
earwax, 196*n*
East Germany, 135
ECoG (electrocorticography), 294–95, 298, 302
École Polytechnique Fédérale de Lausanne (EPFL), 112
ecology, 219
Edison, Thomas, 134, 146
education, 183–84
Edwards, Bradley, 31
EEG (electroencephalogram), 287–90, 291, 292, 294, 298, 299, 310
efficiency, 125–26
eGenesis, 207
EGFRvIII, 243
Egyptians, ancient, 6
Eiben, Gusz, 120*n*
Eiffel Tower, 150, 171
Eisen, Jonathan, 234*n*
electric shock therapy, 299
electromagnetic railgun, 24–25
electrons, 5
Elvis, Martin, 65–67, 68, 320*n*
embryonic stem cells, 273
"emergency guide robot," 130–32
Empire State Building, 172
environment:
   biosynthetic monitoring of, 210–12
   fusion power and, 94
   programmable matter in, 128
   robotic construction and, 155–56
   space flight damage to, 39–40
   synthetic organisms and, 218–19
environmental movement, 97–98
EPFL Laboratory for Timber Construction, 143–44
epilepsy, 295, 302
escape velocity, 55
Escherichia coli, 198

ethanol, 286
Ethnobotany Study Book, 176
European Space Agency (ESA), 22, 27, 65
European Union, 22*n*
Everett, Daniel, 140*n*
evolution, 196
extinction, 221–25
eyes, 186

Faber, Daniel, 53, 68, 69
Fabricated: The New World of 3D Printing (Lipson and Kurman), 159
Fabric of Reality, The (Deutsch), 330
Facebook, 6*n*, 111, 180, 254
face-tracking software, 180
Falcon 9 rocket, 8*n*, 19
Faraday, Michael, 4, 6
fiducial marker, 169–70
fingertips, pruney, 126
Fisher, Caitlin, 173, 177–78
fission, 79*n*
FitBit, 252*n*
flexible electrode arrays, 298
Florida, University of, 300
   Center for Smell and Taste at, 334
Florida State University, 59*n*
flu, 247
flu vaccines, 217
flux pinning, 326–27
flying cars, 2
fMRI (functional magenetic resonance imaging), 290–91
fMRS (functional magnetic resonance spectroscopy), 292
fNIRS (functional near-infrared spectroscopy), 291
food, printed, 159–63
Food and Drug Administration (FDA), 254, 315, 316
foods, 190–91
Ford Motor Company, 97
Forgacs, Gabor, 268–69, 272
forked tongue, 187
"4D printing," 103–5
France, 93*n*
Frankenfood, 221
free fall, 42–43
"freezing of gait," 301
Frostruder, 162
fuel cells, 208–9
fuels, 20, 208–10, 221
furniture, 127
Fusion: The Energy of the Universe (McCracken), 77
fusion bombs, 79
fusion power, 73–100
   benefits of, 93–94
   blast approach to, 84–85
   breakeven point in, 88
   concerns about, 91–93

   confining and heating approach to, 85
   research funding for, 92–93
   where we are now with, 86–90
fusion reactors, 314
Fusor.net, 80
fusors, tabletop, 80–84

Gaia (robot), 129–30
Gatenholm, Dr., 269
gene drive, 201–3
gene expression, 239
General Fusion, 89
genes, 195–96, 197, 204, 215
gene sequencing, 2
Genetic Access Control (app), 251–52
genetic disorders, 3, 219, 235–37
Genetic Information Non-discrimination Act (2008), 250–51
genetic mutations, 40, 231
George Mason University, 56
Georgia Institute of Technology, 130
geostationary orbit, 32, 34, 43
Germany, Nazi, 135
Gilpin, Kyle, 118
GitHub, 251
Global Catastrophic Risks (book of essays), 125*n*
glucose, 286
GMOs (genetically modified organisms), 221
Go-Between, The (Hartley), 331
gold, 52, 92
"Golden Promise" barley, 192
Google, 111, 180, 197*n*, 232, 254, 290
Google Glass, 175–76, 179, 186
Google Scholar, 247
gophers, 96–97
GPS, 171, 173
Gramazio, Fabio, 152
granite, 144
Grant, Dale, 46
gravity, 15–16, 42, 43, 52, 56, 78
"Gray Goo Scenario," 125
gray wolves, 224
Graz University of Technology, 177
Great Britain, 22
Great Depression, 45
Greeks, ancient, 6
Greenpeace, 94*n*
Gunduz, Aysegul, 300, 301
guns, 3D printed, 125

hacking, of brain implants, 309
hands, 323–24, 332
haptic pen, 175
Haque Design + Research (Umbrellium), 111
hard hats, 179
Hartley, L. P., 331
Harvard Medical School, 204, 242

Harvard-Smithsonian Center for
    Astrophysics, 65
Harvard University, 129–30, 150, 203,
    208, 247
Hausbaumaschine, 135
Hayabusa, 65
Hayabusa 2, 65
health care costs, 231
health insurance, 250
hearing, restoring of, 307
heart disease, 246–48
"heat," 5
Heath, Robert, 310
"heavy water," 81
hedgehog grease, 229
Heilig, Morton, 168
helium, 4, 5, 74–75, 76, 91
"High Flight" (Magee), 13
hippocampal prosthesis, 307–8
Hiroshima, 67
Hobbit, The (films), 82
Holocaust, 183
Holocaust Museum, 183
Homeland Security Department, U.S.,
    81
homosexuality, 310
house-building factory, 135
houses:
    programmable, 126
    reconfigurable, 109–11
housing, 134–37, 157
    complexity of, 137
    inspection of, 145–46
Howard Hughes Medical Institute,
    212
Hull, Richard, 80, 82, 84, 92
human genome, 214, 234–35
Human Genome Project, 220
Human Metabolome Database, 244
Huntington's disease, 196*n,* 237
Hussein, Saddam, 48, 49
hydraulic fracturing ("fracking"),
    99
"Hydrocolloid Printing: A Novel
    Platform for Customized Food
    Production" (2009 paper), 162
hydrogen, 4, 73–76, 78, 79, 94, 208–9
hydrogen bombs, 98, 100
hydrogen sulfide, 327
HygroScope, 104
hypertension, 246
hypocholesterolemia, 246
hypothalamus, 189
Hypurin, 198*n*

Ice Age, 223
iGEM (International Genetically
    Engineered Machine), 216
IKEA, 129, 137
Illinois, University of:
    at Urbana-Champaign, 182
    Veterinary School at, 184
Illusio, 185

immune system, 207, 238, 241–42
    organ transplants and, 258–59
immunosuppressive drugs, 258–59, 275
immunotherapy, 242
income distribution, 154
Industrial Revolution, 154
inertial confinement fusion, 86–87
infinite universes, 329, 330
information asymmetry, 181
Innovega, 176
Instagram, 247, 250
Institute of Advanced Architecture of
    Catalonia, 151
insulin, 198, 207
insurance, 250
Interactive Robogami, 108
international arms trade, 48
International Space Station, 15–16, 41,
    42–43
Internet, 109, 122, 216, 262, 269
intracortical neural recording, 295–99
Iowa State University, 179
iPhone, 216
Iraq, 48, 49–50
iron, 54
irritable bowel syndrome, 206
isopropanol, 208–9
isotopes, 73–74
ITER (International Thermonuclear
    Experimental Reactor), 88–89,
    91–94
ivacaftor, 236, 248

Japan, 136
JAXA (Japan Aerospace Exploration
    Agency), 65
J. Craig Venter Institute, 214–15
Jell-O, 298
Jell-O shots, 161
jet fuel, 209–10, 218
Jin, Yaochu, 122
joinery, 143–44
Joint BioEnergy Institute, 210
Joint European Torus (JET), 89
Josephson, Brian, 5–6
Josephson junction, 6
Jurassic Park (film), 222

Kazakhstan, 100
Keasling, Jay, 199
Keating, Steven, 146–48, 153, 155, 253
Kennedy, Philip, 315–17
Kevlar, 35
Khoshnevis, Behrokh, 145, 146, 147,
    158
kidneys (organ), 280
Kilobot project, 115, 119
Kohler, Matthias, 152
Kurman, Melba, 159

Lake Chagan, 100
lasers, 2, 27–29, 84, 86–87
Law of the Sea, 33

leukemia, 238, 239, 242
Leuthardt, Eric, 303, 314–15
Levin, Gilbert, 334
levitation, 326–27
LiDAR, 174
life insurance, 250
LIFT (laser-induced forward transfer),
    265–66
Limited Test Ban Treaty (LTBT), 99
Lipschultz, Bruce, 91–92, 93
Lipson, Hod, 159
Lipton, Jeffrey, 162
liquid hydrogen, 39
liquid oxygen, 20, 39
lithium, 77
LIT ROOM, 110–11
livers (organs), 257–59, 260–61, 280
lizards, 187
locked-in syndrome, 316
Lockheed Martin, 90
lossless power transmission, 325
Low Earth Orbit (LEO), 14, 15–16,
    21, 34, 38
Lowther, William, 50
lung cancer, 238–40
lungs (organ), 261
Lyme disease, 255
lymphoma, 242

McAlpine, Michael, 271
McCracken, Garry, 77
Magee, John Gillespie, Jr., 13
"magic book," 176
MagLIF (Magnetized Liner Inertial
    Fusion) project, 87–88
"magnetic confinement"-type reactors,
    85
magnetic levitation (MagLev) trains,
    24–25, 30, 327
magnetosphere, 59
magnets, 5
MakerBot, 162
malaria, 198–203, 207
mammoth genome, 222–24
Mankins, John, 320
marble, 144
marching bands, 119–20
Mars, 19, 40, 45*n,* 52, 55, 158–59
Mars One project, 45*n*
Masiello, Carrie, 210–11
Massachusetts General Hospital,
    242
Massachusetts Institute of Technology
    (MIT), 102, 103, 104, 106, 107*n,*
    108, 214, 216
    Mediated Matter lab at, 146
    Plasma Science and Fusion Center
    at, 91
matching markets, 275–81
Matthews, Kirstin, 250
Maus, Marcela, 242–43
Max Planck Institute for Infection
    Biology, 212

INDEX

μBiome, 2
M-blocks, 118
MD Anderson Cancer Center, 232, 234
Mediated Matter lab, 146
medical tourism, 272
medical trials, 254–55, 268–69
medicine, 221
  augmented reality in, 179, 185–86
  bioprinting and, *see* bioprinting
  origami robots in, 106–7
  programmable matter in, 127–28
  synthetic biology in, 198–207
  see also precision medicine
Meetup.com, 175, 179
MEG (magnetoencephalography), 289–90, 291
Meissner effect, 326
meltdown, 91–92
memory, 220, 304, 307–8, 311
Mendelsohn, John, 232, 234
Meng, Yan, 122
Menges, Achim, 104
Menon, Sandeep, 235
messenger RNA, 193
metabolome, 244–46
meteorites, 53, 67
Michigan Array, 296, 298
microRNA, 239–40, 246–47
Microsoft, 272
Miller, Jordan, 261, 269, 270–71, 274
miniaturization, 176
"Minibuilders," 151–52
miRBase, 240
mirror humans, 332–35
MIT Technology Review, 6n
molds, configurable, 134
molecular scissors, 212, 213–14
molecules, mirror, 334
monogenic traits, 196–97
mononucleosis, 230
moon, 55
moon landing, 19
moral hazard, 273–74
Moravec's Paradox, 139
mosquitoes, 200, 203, 218
Mossad, 50
motion sickness, 168
movies, 183
MRI (magnetic resonance imaging), 290–91
M-type (metal) asteroids, 53, 54
mucociliary escalator, 187–88
mucus, 236
Mukhopadhyay, Aindrila, 210
multiverse, 329
Munger, Steven, 334–35
Musk, Elon, 19
mutation breeding, 191–92
mutations, 219, 236–37
Mycoplasma genitalium, 214–15
Mycoplasma laboratorium, 215
Mycoplasma mycoides, 215n

Nagasaki bombing, 98
nano-bio-machines, 3
nanobots, 118
nanotechnology, 221
NASA Innovative Advanced Concepts (NIAC), 25, 31, 35
nasal cycle, 186–89
nasal venous sinusoids, 188
NASA (National Aeronautics and Space Administration), 20, 47, 60, 65, 92, 158, 159–60
National Academy of Sciences, 203
National Cancer Institute, 238
National Defence Department, Canada, 47
National Ignition Facility (NIF), 86–87
National Institutes of Health, 214, 234, 235
Native Americans, 196n
natural gas, 73, 98–99
Nebraska, University of, 176
Neufert, Ernst, 135
neural dust, 299
neural implants, 310
Neurobridge, 312
neuro-cyber-connection, 312–13
neurons, 286–87, 290, 298, 306
  EEGs and, 287–90
NeuroPace, 302
neuroprosthetics, 311, 315, 322, 324
neurotrophic electrodes, 297–98, 315, 316
Neutron Club, 80
neutron gun, 80–81
neutrons, 73, 91
New Jersey, 299
New Mexico, 96
nickel, 54
Nocera, Dan, 208
North Carolina State University, 63
Norway, 22n
nostrils, 186–89
Nuclear Explosions for the National Economy, 100
nuclear reactors, 58
Nucleon (concept car design), 97
nucleus, 192, 193
nutrition, 245–46

Olestra, 334
Oliver, John, 326n
Open Humans Foundation, 252n
"optical mining," 63
orbiting factory, 24
organ donation, 257n
organ markets, 274, 275–81
Organovo, 268
organ rejections, 275
organ sales, 258, 280–81
organ transplant list, 257–58, 272
organ transplants, 206–7
origami robots, 105–8, 129

OSIRIS-REx, 65
"Our Friend the Atom" (Disney cartoon), 97
Outer Space Treaty (1967), 63–64
oxidizer, 20
Oxman, Neri, 146, 148
oxygen, 208–9
oxygen deprivation, 205
oxygen gas, 82

Pacific Ocean, 35–36
Paddon, Chris, 199
Palo Alto Research Center, 116
Panama Canal, 97
pancreas, 236
parallel universe, 329
paralysis, 312
Parkinson's disease, 301
patenting, 124
patent law, 272
peacekeepers, 181
Pennsylvania, University of, 108
Personal Genome Project, 252–53
personal security, 124–25
PERVs, 207
pesticides, 200
Petersen, Kirstin, 149, 150–51
Pfizer, 235
phobias, 179
Phobos (moon of Mars), 55
phosphenes, 306
photosynthesis, 208
Picon, Antoine, 138
pigs, 206
Piraha (Amazonian tribe), 140n
Pitt, Brad, 167
Plait, Phil, 36, 38
plants, 125
  Chinese sweet wormwood, 198–99
plasma, 85, 88
Plasma Science and Fusion Center, 91
platinum, 52, 55
pluripotent stem cells, 273
plutonium, 58
pogo sticks, 27
Pokémon GO, 8n, 166, 182–83
pollution, 94
porcine endogenous retroviruses (PERVs), 207
positive transcriptional autoregulation, 205n
potassium iodide pills, 60
poverty, 157
precision medicine, 229–56
  benefits of, 254–56
  cancer diagnosis, treatment, and monitoring in, 238–44
  concerns about, 248–53
  data collection in, 234–35
  genetic disorders and, 235–37
  metabolome and, 244–46
  privacy issues in, 248, 250–53

Precision Medicine Initiative Cohort Program, 234
predictive ability, 1–2
Princeton University, 142, 271
privacy issues, 130, 182, 248
  of AR, 180–81
  in brain-computer interfaces, 309–10
  in precision medicine, 248, 250–53
programmable matter, 101–32
  benefits of, 125–29
  computers as, 101
  concerns about, 122–25
  in everyday life, 105
  hacking of, 122–23
  military applications of, 123–24
  origami robots as, 105–8
  power for, 118
  reconfigurable houses and, 109–11
  see also robots
programmed materials, 103–5
Project Babylon, 48–49
Project Esper, 185
Project HARP (High Altitude Research Project), 47, 48
Project Plowshare, 96–100
Project Rulison, 98
Promobot, 129
Promobot IR77, 129
propellants, 14–15, 18, 20, 23
prostate cancers, 239n, 247
prosthetics, advanced, 322–24
proteins, 193, 194, 195, 221, 234, 239, 332
protium, 73
protons, 73, 77
Pryor, Richard, 328n

QR code, 169–71
quantum computing, 328–30
quantum mechanics, 329, 330
Quinn, Roger, 151n

radiation, 59–60, 62, 99
radiation therapy, 241
radioactive waste, 91
railgun, electromagnetic, 24–25
ramjet, 21, 22, 26
Reaction Engines, 22
Recognizer, 180
Reconfigurable House exhibit, 111
recycled fecal matter, 160
recycling, 128
Reece, Andrew, 247
refining, 56
refrigeration, 4
"Registry of Standard Biology Parts," 216
Reichert, Steffen, 104
Reiss, Louise and Eric, 99
RepRap, 269–70
"repugnance," in markets, 276
reuse, 128
ribosome, 193–94, 195

Rice University, 200n, 210, 250, 261
rigid airship, 29–30
Ringeisen, Bradley, 259
RNA, 193–94, 195, 332
RNS System, 302
Robinette, Paul, 130
Robot Baby Project, 120n
robotic construction, 134–63
  benefits of, 156–59
  concerns about, 153–56
  and space travel, 158–59
  swarm robots in, 149–53
  3D printing for, 144–49
robots, 102, 129–32
  autonomous, 113–16
  as construction workers, 139–44
  coordinating movement of many, 119–22
  evolving of, 120–22
  generalization in, 142
  industrial, 136
  in medicine, 127–28
  modular, 112–16
  neuroprosthetics and, 311
  origami, 105–8, 129
  termite-inspired, 150–51
  see also programmable matter
rocket launches, 3
rockets, 23, 39
  air-breathing, 19–24
  aircraft-launched, 29–30
  cost of, 14
  laser ignition for, 27–29
  propellant for, 14–15, 18, 20, 23
  reusable, 14, 15, 18–19, 39
  simplicity of, 22
  stages of, 18n
rocket sled, 25, 26
rockoon, 29
rod from God, 38
roller coaster, 23, 42
Romanishin, John, 118
Roombots, 112–13, 121, 127
Roth, Alvin, 276, 277, 279, 280
"Ruby Red" grapefruit, 192
Rus, Daniela, 106–7, 108, 118, 128
Russia, 67, 99, 217n

SABRE (Synergetic Air-Breathing Rocket Engine), 22
Saddest Generation, 166
Safe Is Not an Option: Overcoming the Futile Obsession with Getting Everyone Back Alive That Is Killing Our Expansion into Space (Simberg), 44
Sahara Desert, 321
SAM (robot), 141, 142, 153–54
Sandia Labs, 85, 87
San Francisco, Calif., 154
sanitation, 157
satellites, 20, 34, 41, 47
Schalk, Gerwin, 313

Schall, Gerhard, 177
Schrödinger's cat, 329
Schrödinger's Killer App (Dowling), 330n
Schwenk, Kurt, 187
See No Evil, Hear No Evil (film), 328n
seizures, 300, 301, 302
Select Sires, Incorporated, 197n
self-driving cars, 123
Sensorama, 168
Shapiro, Beth, 222, 223–24
Shotwell, Gwynne, 19
Shtetl-Optimized (blog), 330n
Siberia, 224
sickle cell amenia, 237
Silberg, Joff, 210–11, 218–19
silicon, 52, 54
Silver, Pamela, 204, 205–6, 208–10, 219
Simberg, Rand, 44
Skylon, 22
Skype, 314
Skywalker, Luke (char.), 324
Slingatron, 25–26
slums, 157
smallpox, 216, 217
Smart Helmet, 179
"smart homes," 111
smartphones, 169
smell, sense of, 174–75, 186–89, 334
Smith, Noah, 153n, 154
snakes, 187
social media, 248, 250
  privacy issues of, 180–81
software, 102, 104–5, 124
  hacking of, 122
solar flares, 60
solar panels, 58
  cost of, 320
solar photovoltaic cells, 92, 208
solar power, space-based, 319–21
solar wind, 37
Solid Freeform Fabrication Symposium, 162
solid rocket boosters, 39
solid tumors, 238, 240–41
Solomon, Scott, 200n
sound, speed of, 21
South Africa, 48
Southern California, University of, 145, 308
Soviet Union, 38, 58, 99, 100, 135
space cannon, 23–26
space debris, 39–40
space elevators, 31–38, 39, 41, 42–43, 314, 320
spaceflight, 13–50
  air-breathing rockets and spaceplanes for, 19–24
  benefits of, 41–45
  concerns about, 38–40
  cost of, 41, 44–45

present cost of, 13–14
reusable rockets for, 18–19
space elevators and tethers for, 31–38
starting at high altitude, 29–30
spaceplanes, 19–24, 39
space settlements, 40
Space Shuttle, U.S., 18, 39
space tethers, 31–38
space tourism, 42
space travel:
  fusion energy in, 94
  supergun for, 23–26
SpaceX, 8n, 18, 19, 30
spatial resolution, 288, 289, 292–93
spearmint, 334
spinal damage, 312
Sputnik, 39
SR-71 spy plane, 21
Starbucks, 180
Star Trek franchise, 34, 86
Star Wars franchise, 78n, 82, 263
steam turbine, 76
stem cells, 263, 272–73
Stevens Institute of Technology, 92, 122
STL-file, 267
storytelling, 178
stratospheric spaceport, 29–30
straw, reconfigurable, 103–4
stress, 246
stroke, 317
strong nuclear force, 77
strontium-90 (Sr-90), 99
Stuttgart, University of, 104, 143
S-type (stony) asteroids, 53, 54
sugar molecules, 210
sugar sintering, 270–71
sun, 59, 78
Sung, Cynthia, 108, 119, 127
superconducting levitation, 326–27
superconducting quantum interference
  device (SQUID), 4, 6, 290
superconductors, 4–6
  room-temperature, 325–28
supergun, 46–50
supersonic ramjet ("scramjet"), 21–22, 26, 126
Sure Shot Cattle Company, 197n
surgery, 185–86
Surrey, University of, 122
swarm bots, 119–20, 121–22
SWARMORPH project, 113–15
swarm robots, 149–53
switchgrass, 209–10
Switzerland, 22n
SYMBRION, 115
Syn 3.0, 215
synthetic biology, 190–225
  benefits of, 220–21
  concerns about, 216–19
  environmental monitoring by, 210–12
  fuel production by, 208–10

generalizing of, 212–14
grassroots approach to, 216
"Synthetic Biology for Recycling
  Human Waste into Food,
  Nutraceuticals, and Materials:
  Closing the Loop for Long-Term
  Space Travel" project, 160
synthetic materials, 101–2
syphilis, 230n
Syria, 156
Systems & Materials Research
  Consultancy, 159

T cells, 242–43
technology, 3–4
  asteroid moving, 67
  contingent nature of development of, 3–7
  discontinuous leaps in, 2
Telegraph, 183
Teller, Edward, 98
temporal resolution, 288, 292–93
Terminator (film), 103
termites, 120, 149, 150–51
terrorism, 36, 38, 217
Tethers Unlimited, 63
tetracycline, 200
theft, 130
3D printers, 144–49, 151–52, 259
  prosthetics and, 322
3D printing, 125, 152
  of food, 159–63
  of organs, *see* bioprinting
  software for, 267
3554 Amun, 53
Throw Trucks with Your Mind (game), 312
thyroid, 60
Tibbits, Skylar, 103–5, 118, 123, 126
titanium, 35
"tokamak" configuration, 88, 92
tornados, 25
touch, sense of, 175
Tourette's syndrome, 301
transcranial magnetic stimulation, 302, 304
transfer RNA, 193–94, 195
Transformers series, 102
The Tree of Life (Web site), 234n
tritium, 74, 77n, 91
tumor cells, 205
tumors, 290
"Tunable Protein Piston That Breaks
  Membranes to Release
  Encapsulated Cargo, A" (Silver, et
  al.), 206
"Tunguska event" (1908), 67
turbofan engine, 20–21, 22
Turner, Ron, 35, 36, 37
23andMe, 251, 252
Twitter, 20n, 187, 250
Two and a Half Men (TV show), 310
Type II superconductors, 326

Umbrellium (Haque Design +
  Research), 111
Underground Railroad, 178
UN-Habitat, 157
Unilateral Forced Nostril Breathing
  (UFNB), 189
United Nations, 96
United States, 39, 135–36
Universal Semen Sales, Inc., 197n
uranium, 58
U.S. Commercial Space Launch
  Competitiveness Act (2015), 64
Utah Array, 295–96, 298, 312

van Pelt, Michel, 27, 37, 39, 40
vasculature, 262
Velvet Glove (Canadian missile
  program), 45
Venter, J. Craig, 214–15
Ventura, Jonathan, 176–77
Vermont, University of, 247
Vietnam War, 99
Virgin Galactic, 30
"virtual mirror interface," 177
virtual reality (VR), 165, 168–69, 182
  headsets for, 168, 169
  motion sickness in, 168
viruses, 212–13
Vishik, Inna, 328
vision, restoring of, 306–7
vomeronasal organs, 107

Wadsworth Center, 313
Wagner, Daniel, 250, 251
Wang, Caddie, 269
warfare, 181
Washington University, 303
water, 53, 60, 81, 208–9
water memory, 5
Weaver, Sigourney, 83
Weightless Rendezvous And Net
  Grapple to Limit Excess Rotation
  (WRANGLER) System, 63
Wellerstein, Alex, 83n, 92
Werfel, Justin, 150–51
Whedon, Joss, 78n
Whegs, 151
White, David, 188, 189
white blood cells, 238, 242
White House, 96
Who Gets What And Why (Roth), 276
Wikipedia, 6n, 319
Wilder, Gene, 328n
Wilderness of Mirrors (Grant), 46
Wilson, Taylor, 80
Winfield, Alan, 115n, 120n
WinSun, 146
Wolverine, 83
wolves, 191, 224
wood joints, 143–44
woodworking, 142–44
workplace safety, 309

World Health Organization, 198
World War II, 135

xenotransplantation, 206–7

Yang, Luhan, 207
yeast, 197, 207

Yellowstone National Park, 224
York, University of, 91
York University, Augmented Reality
    Lab at, 173, 177–78
YouTube, 141
Yuri (unit of specific strength),
    35

Zhang, Feng, 214
Zimov, Sergey, 224
Z machine, 85, 88
Zoloft, 180
Zurich, 152